COMPUTER MATHEMATICS

COMPUTER MATHEMATICS

Gerard Prudhomme

ARCLER
PRESS

www.arclerpress.com

Computer Mathematics

Gerard Prudhomme

Arcler Press
2010 Winston Park Drive,
2nd Floor
Oakville, ON L6H 5R7
Canada
www.arclerpress.com
Tel: 001-289-291-7705
001-905-616-2116
Fax: 001-289-291-7601
Email: orders@arclereducation.com

ISBN: 978-1-77361-405-2 (Hardcover)

ABOUT THE AUTHOR

Gerard I. Prudhomme has a graduate degree (M.S.) for Computer Science from University College London (UCL). He has also worked as a software programmer and tech writer for different Fortune 500 companies, and studied at UCL, Harvard, and Oxford.

TABLE OF CONTENTS

LIST OF FIGURES

https://creativecommons.org/licenses/by-sa/3.0/deed.en
Author: Wvbailey

Figure 6.7. Flowchart of an algorithm
Creative Commons Attribution-Share Alike 4.0 International
https://creativecommons.org/licenses/by-sa/4.0/deed.en
Author: Somepics

Figure 6.8. Logical NAND algorithm
Attribution-ShareAlike 3.0 Unported (CC BY-SA 3.0)
https://creativecommons.org/licenses/by-sa/3.0/deed.en
Author: Sakurambo

Figure 6.9. Böhm-Jacopini structures
Attribution-ShareAlike 3.0 Unported (CC BY-SA 3.0)
https://creativecommons.org/licenses/by-sa/3.0/deed.en
Author: Wvbailey

Figure 6.10. Euclid's algorithm Book VII
Attribution-ShareAlike 3.0 Unported (CC BY-SA 3.0)
https://creativecommons.org/licenses/by-sa/3.0/deed.en
Authors: Wvbailey

Figure 10.1. Knuth's version of Euclid's algorithm
Creative Commons Attribution-Share Alike 3.0 Unported license.
https://creativecommons.org/licenses/by-sa/3.0/deed.en
Author: Wvbailey

Figure 10.2. Turing machine
Creative Commons Attribution-Share Alike 3.0 Unported license.
https://creativecommons.org/licenses/by-sa/3.0/deed.en
Author: Wvbailey

Figure 10.3. Two differential areas
Creative Commons Attribution-Share Alike 3.0 Unported license.
https://creativecommons.org/licenses/by-sa/3.0/deed.en
Author: Dov Grobgeld

Figure 10.4. Turing machine
Creative Commons Attribution-Share Alike 3.0 Unported
https://creativecommons.org/licenses/by-sa/3.0/deed.en
Author: Wvbailey

Figure 10.5. State diagram
Creative Commons Attribution-Share Alike 3.0 Unported
https://creativecommons.org/licenses/by-sa/3.0/deed.en
Author: Wvbailey

Figure 10.6. 3-state P-T machine
Creative Commons Attribution-Share Alike 3.0 Unported
https://creativecommons.org/licenses/by-sa/3.0/deed.en
Author: Wvbailey

Figure 10.7. 2-state P-T machine
Creative Commons Attribution-Share Alike 3.0 Unported
https://creativecommons.org/licenses/by-sa/3.0/deed.en
Author: Wvbailey

Figure 10.8. Turing machine
Creative Commons Attribution-Share Alike 3.0 Unported
https://creativecommons.org/licenses/by-sa/3.0/deed.en
Author: Wvbailey

Figure 10.9. Universal turing machine
Creative Commons Attribution-Share Alike 3.0 Unported

https://creativecommons.org/licenses/by-sa/3.0/deed.en
Author: Cbuckley

Figure 10.10. Godel machine
Creative Commons Attribution-Share Alike 3.0 Unported
https://creativecommons.org/licenses/by-sa/3.0/deed.en
Author: Wvbailey

Figure 10.11. Diagram 3 states
Creative Commons Attribution-Share Alike 3.0 Unported
https://creativecommons.org/licenses/by-sa/3.0/deed.en
Author: Wvbailey

Figure 14.1. State diagram 3 states
Creative Commons Attribution-Share Alike 3.0 Unported
https://creativecommons.org/licenses/by-sa/3.0/deed.en
Author: Diego Queiroz

Figure 14.2. Automata theory
Creative Commons Attribution-Share Alike 4.0 International
https://creativecommons.org/licenses/by-sa/4.0/deed.en
Author: ManiacParisien

Figure 14.3. Turing machine
Creative Commons Attribution-Share Alike 3.0 Unported
https://creativecommons.org/licenses/by-sa/3.0/deed.en
Author: Nynexman4464 at the English language Wikipedia

Figure 14.4. Oracle machine
Creative Commons Attribution-Share Alike 4.0 International
https://creativecommons.org/licenses/by-sa/4.0/deed.en
Author: Fschwarzentruber

Figure 14.5. Turing machine
Creative Commons Attribution-Share Alike 3.0 Unported
https://creativecommons.org/licenses/by-sa/3.0/deed.en
Author: RosarioVanTulpe

Figure 14.6. 2 + 3 = 5 Turing machine
Creative Commons Attribution-Share Alike 3.0 Unported
https://creativecommons.org/licenses/by-sa/3.0/deed.en
Author: BenTels at nl.wikipedia

Figure 14.7. Diagonalizing function
Creative Commons Attribution-Share Alike 3.0 Unported
https://creativecommons.org/licenses/by-sa/3.0/deed.en
Author: Wvbailey

Figure 14.8. Diagonalizing machine
Creative Commons Attribution-Share Alike 3.0 Unported
https://creativecommons.org/licenses/by-sa/3.0/deed.en
Author: Wvbailey

Figure 14.9. UML Object
Version 3 of the GNU General Public License
https://www.gnu.org/copyleft/gpl-3.0.html
Author: Telecomtom

Figure 14.10. Theoretical computer science
Creative Commons Attribution-Share Alike 3.0 Unported
https://creativecommons.org/licenses/by-sa/3.0/deed.en
Author: RobinK

Figure 14.11. Turing Test
Version 3 of the GNU General Public License
https://www.gnu.org/copyleft/gpl-3.0.html
Author: Hugo Férée

Figure 16.1. KL Gauss example.
Creative Commons Attribution-Share Alike 3.0 Unported.
https://creativecommons.org/licenses/by-sa/3.0/deed.en.
Author: T. Nathan Mundhenk.

Figure 16.2. Cauchy distributions.
Creative Commons Attribution-Share Alike 3.0 Unported.
https://creativecommons.org/licenses/by-sa/3.0/deed.en.
Author: Skbkekas.

Figure 16.3. Logarithm and illustration of the derivative.
Creative Commons Attribution-Share Alike 3.0 Unported.
https://creativecommons.org/licenses/by-sa/3.0/deed.en.
Author: Krishnavedala.

Figure 16.4. Log–log plot.
Creative Commons Attribution-Share Alike 4.0 International.
https://creativecommons.org/licenses/by-sa/4.0/deed.en.
Author: M. W. Toews.

Figure 16.5. Law of the wall.
Creative Commons Attribution-Share Alike 3.0 Unported.
https://creativecommons.org/licenses/by-sa/3.0/deed.en.
Author: aokomoriuta.

Figure 16.6. PISO algorithm.
Creative Commons Attribution-Share Alike 3.0 Unported.

https://creativecommons.org/licenses/by-sa/3.0/deed.en.
Author: Iit.asheesh.

Figure 16.7. Evolutionary algorithm.
Creative Commons Attribution-Share Alike 3.0 Unported.
https://creativecommons.org/licenses/by-sa/3.0/deed.en.
Author: Jorge.maturana.

Figure 16.8. Evolutionary algorithm
Creative Commons Attribution-Share Alike 3.0 Unported
https://creativecommons.org/licenses/by-sa/3.0/deed.en
Author: Johann "nojhan" Dréo

Figure 16.9. Error reduction algorithm
Creative Commons Attribution 2.5 License
https://creativecommons.org/licenses/by/2.5/
Authors: Morner

Figure 16.10. Diagnostic algorithm
Creative Commons Attribution-Share Alike 3.0 Unported
https://creativecommons.org/licenses/by-sa/3.0/deed.en
Authors: KimHasselfeld

Figure 16.11. Cryptography algorithm
Creative Commons Attribution-Share Alike 3.0 Unported
https://creativecommons.org/licenses/by-sa/3.0/deed.en
Author: Mercurish

Figure 16.12. International data encryption algorithm
Creative Commons Attribution-Share Alike 3.0 Unported
https://creativecommons.org/licenses/by-sa/3.0/deed.en
Author: Surachit

Figure 18.1. Algorithm for diagnosis of heart failure
Creative Commons Attribution-Share Alike 3.0 Unported
https://creativecommons.org/licenses/by-sa/3.0/deed.en
Author: InfoHEART

Figure 18.2. Imperialist competitive algorithm
Creative Commons Attribution-Share Alike 3.0 Unported
https://creativecommons.org/licenses/by-sa/3.0/deed.en
Author: Icasite

Figure 18.3. Csmaca algorithm
Creative Commons Attribution-Share Alike 3.0 Unported
https://creativecommons.org/licenses/by-sa/3.0/deed.en
Author: Tkz4

Figure 18.4. An algorithm to find functional DNA methylation in cancer cells
Creative Commons Attribution-Share Alike 4.0 International
https://creativecommons.org/licenses/by-sa/4.0/deed.en
Author: Ssridhar17

Figure 18.5. ID3 algorithm
Creative Commons Attribution-Share Alike 4.0 International
https://creativecommons.org/licenses/by-sa/4.0/deed.en
Author: Acoggins38

Figure 18.6. K-means convergence
Creative Commons Attribution-Share Alike 3.0 Unported
https://creativecommons.org/licenses/by-sa/3.0/deed.en
Author: Chire

Figure 21.1. K-means Gaussian
Creative Commons Attribution-Share Alike 3.0 Unported
https://creativecommons.org/licenses/by-sa/3.0/deed.en
Author: Chire

Figure 21.2. Hierarchical algorithm
Creative Commons Attribution-Share Alike 3.0 Unported
https://creativecommons.org/licenses/by-sa/3.0/deed.en
Author: Victorbabkov

Figure 21.3. Quick-Search algorithm
Creative Commons Attribution-Share Alike 3.0 Unported
https://creativecommons.org/licenses/by-sa/3.0/deed.en
Author: Scl758

Figure 21.4. Algorithm diagram for XXTEA
Creative Commons Attribution-Share Alike 3.0 Unported
https://creativecommons.org/licenses/by-sa/3.0/deed.en
Author: Alois Mahdal

Figure 21.5. Tomasulo's algorithm
Creative Commons Attribution-Share Alike 4.0 International
https://creativecommons.org/licenses/by-sa/4.0/deed.en
Author: Tylerc0726

Figure 23.1. Boruvka's algorithm
Creative Commons Attribution-Share Alike 3.0 Unported
https://creativecommons.org/licenses/by-sa/3.0/deed.en
Author: Swfung8

Figure 23.2. Breadth-first-search-algorithm
Creative Commons Attribution-Share Alike 3.0 Unported

https://creativecommons.org/licenses/by-sa/3.0/deed.en
Author: Mre

Figure 23.3. Traffic sign detection algorithm
Creative Commons Attribution-Share Alike 4.0 International
https://creativecommons.org/licenses/by-sa/4.0/deed.en
Author: Vidyakv

Figure 23.4. COMP algorithm
Creative Commons Attribution-Share Alike 4.0 International
https://creativecommons.org/licenses/by-sa/4.0/deed.en
Author: CheCheDaWaff

Figure 23.5. Genetic algorithm
Creative Commons Attribution-Share Alike 3.0 Unported
https://creativecommons.org/licenses/by-sa/3.0/deed.en
Author: Yearofthedragon

Figure 23.6. Tortoise and hare algorithm
Creative Commons Attribution-Share Alike 3.0 Unported
https://creativecommons.org/licenses/by-sa/3.0/deed.en
Author: David Eppstein

Figure 23.7. Chan algorithm
Creative Commons Attribution-Share Alike 3.0 Unported
https://creativecommons.org/licenses/by-sa/3.0/deed.en
Author: Sergey539

Figure 23.8. Dijkstra algorithm
Creative Commons Attribution-Share Alike 3.0 Unported
https://creativecommons.org/licenses/by-sa/3.0/deed.en
Author: Willprice

Figure 23.9. Algorithm for dynamic smile reconstruction
Creative Commons Attribution-Share Alike 3.0 Unported
https://creativecommons.org/licenses/by-sa/3.0/deed.en
Author: Elyse Verboom

Figure 23.9. Gift wrapping algorithm
Creative Commons Attribution-Share Alike 4.0 International
https://creativecommons.org/licenses/by-sa/4.0/deed.en
Author: Maonus

Figure 23.10. Critical path algorithm
Creative Commons Attribution-Share Alike 3.0 Unported
https://creativecommons.org/licenses/by-sa/3.0/deed.en
Author: Petr Kopač

Figure 23.11. FSA-Red Algorithm
Creative Commons Attribution-Share Alike 3.0 Unported
https://creativecommons.org/licenses/by-sa/3.0/deed.en
Author: Ferisulianta

Figure 23.12. Matrix multiplication example
Creative Commons Attribution-Share Alike 3.0 Unported
https://creativecommons.org/licenses/by-sa/3.0/deed.en
Author: Ayush3504

Figure 23.13. Tiled matrix multiplication diagram
Creative Commons Attribution-Share Alike 3.0 Unported
https://creativecommons.org/licenses/by-sa/3.0/deed.en
Author: Ayush3504

Figure 26.1. Chord routing algorithm
Creative Commons Attribution-Share Alike 3.0 Unported

https://creativecommons.org/licenses/by-sa/3.0/deed.en
Author: Mycure

Figure 26.2. Algorithm cancer magnetic nanotherapy
Creative Commons Attribution-Share Alike 4.0 International
https://creativecommons.org/licenses/by-sa/4.0/deed.en
Author: V-orel

Figure 26.3. FPF algorithm
Creative Commons Attribution-Share Alike 3.0 Unported
https://creativecommons.org/licenses/by-sa/3.0/deed.en
Author: Skhakabi

Figure 26.4. ALS algorithm
Creative Commons Attribution-Share Alike 4.0 International
https://creativecommons.org/licenses/by-sa/4.0/deed.en
Author: Jfdwolff

Figure 26.5. Molecular dynamics algorithm
Creative Commons Attribution-Share Alike 3.0 Unported
https://creativecommons.org/licenses/by-sa/3.0/deed.en
Author: Knordlun

Figure 26.6. Boruvka's (Sollin's) algorithm
Creative Commons Attribution-Share Alike 3.0 Unported
https://creativecommons.org/licenses/by-sa/3.0/deed.en
Author: Alieseraj

Figure 26.7. Backward algorithm – Hidden Markov models
Creative Commons Attribution-Share Alike 3.0 Unported
https://creativecommons.org/licenses/by-sa/3.0/deed.en
Author: Enrique Benimeli

Figure 26.8. Scarf's algorithm
Creative Commons Attribution-Share Alike 3.0 Unported
https://creativecommons.org/licenses/by-sa/3.0/deed.en
Author: Zyang666

Figure 26.9. Yen's K-shortest path algorithm
Creative Commons Attribution-Share Alike 3.0 Unported
https://creativecommons.org/licenses/by-sa/3.0/deed.en
Author: KRPent

Figure 26.10. Prim's algorithm
Creative Commons Attribution-Share Alike 3.0 Unported
https://creativecommons.org/licenses/by-sa/3.0/deed.en
Author: fungszewai

Figure 26.11. Smith-Waterman-algorithm
Creative Commons Attribution-Share Alike 4.0 International
https://creativecommons.org/licenses/by-sa/4.0/deed.en
Author: Yz cs5160

Figure 26.12. Genetic algorithm crossover
Creative Commons Attribution-Share Alike 3.0 Unported
https://creativecommons.org/licenses/by-sa/3.0/deed.en
Author: Yearofthedragon

Figure 26.13. Shell sorting algorithm
Creative Commons Attribution-Share Alike 4.0 International
https://creativecommons.org/licenses/by-sa/4.0/deed.en
Author: Balu Ertl

Figure 26.14. Marching squares algorithm
Creative Commons Attribution-Share Alike 4.0 International

PREFACE

The title "Computer Mathematics" deals with a plethora of concepts and covers everything from computer science, to answer mathematical questions, technical mathematical aspects, playing the role of a committee for proposed mathematical and computer science activities, the scientific management of a mathematical and computer science site, mobilizing competent mathematical and computer science staff, mobilizing a team of mathematical and computer science consultants, etc. The flexibility of computer science makes it possible to also choose more varied examples: daily digital devices, sound, robots, physical simulations, video images, textual data, etc.

On the other hand, we may also, to provide an example, use a finite element method (FEM), which allows numerical computation of an approximate solution. To illustrate, we may additionally supply mathematical modeling, and with a mathematical model accounting for the physical laws to which all structures involved (biological and medical laws, mechanical and physical laws, etc.) obey.

It is interesting to note that these fundamental trends still constitute the basis of contemporary computing. The meetings therefore held periodically by mathematicians who teach in engineering and computer science colleges have the purpose and effect of getting them out of isolation. In this book, we will present the rudiments of this discipline and the classical results of complexities of the descriptive and logical descriptions of the classes of complexity.

Some may feel the teaching of math is a waste of time for almost all those who suffer it. On the other hand, for an example of educational mathematical and computer science studies, the "Computer Science Unplugged" educational projects advocates teaching computer science with strings, maps, and pencils rather than distributing digital tablets in colleges. So to oppose mathematical and computer science studies are misunderstanding, that is in any case in my opinion, a little the main subject of computer science, that is: to experiment, discover, learn, try, search, etc.

After the computer, the network is thus the second instance of this machine concept, which gives this question of extension in space an essential place.

Thus, mathematical and computer science studies are based on concepts, to provide examples: the shape of electrolytic cells for aluminum on mathematical analysis, the analysis of images on a recent generalization of the Fourier analysis, the transmission of messages on algebraic combinatorics, integral calculation results, etc.

Admittedly, users of mathematical and computer science studies showing interest in algorithms is not solely inspired by the advent of the computer. To illustrate, we may additionally supply computer science studies of a site on the web as a center of mathematical consultation and resources. Living matter, inanimate matter, and the networks at work in human relations all constitute the general frameworks of the mathematical, physical, and human sciences. Some may say we give too much importance to math for a computer science curriculum info, however, they are wrong.

We should all, certainly, therefore be sensitive to this and the diversity of machines. A machine is a tool, that is to say a material system, which obeys the laws of physics. For colleges, where the physical sciences are not explicitly listed, we recommend that physicists and chemists give at least their opinion on mathematics programs.

This logical feasibility thus reveals a profound analogy between the programming of computer algorithms and the development of mathematical demonstrations. Finally, this book will introduce an introduction to the complexity of constraint satisfaction problems. You will also learn about the history of mathematics.

And, we will also find a very useful consideration of societal aspects of the use of digital machines. Remember, mathematical demonstrations require the elementary levels of the physical sciences, and also probabilities and the notion of variability. It is, therefore, a challenge to identify a set of general methods and to find what might be the analog of conservation laws in physics or the second principle of thermodynamics for computation.

In parallel, at the turn of the twentieth century, the axiomatic current conquers many branches of mathematics, with as a corollary metamathematics or methodological questions giving rise to a new discipline: mathematical logic. Mental arithmetic and calculating orders of magnitude should therefore find new reasons to practice mathematical and computer techniques.

1
CHAPTER

PREPARATORY MATHEMATICAL AND COMPUTER SCIENCE STUDIES

Therefore, after emphasizing the articulation of mathematics as a science with other disciplines, we need to examine how this is reflected in teaching, and make suggestions for improving the situation. This point of view is inadequate for a computer scientist, for whom the abstract functions as an algorithm.

On the other hand, it would be ridiculous to end a mathematics book at the preparatory course level because the course has not yet dealt with numeration. To illustrate, we may additionally supply the fact that computers are endowed with a powerful mathematical device that guides the design of robust, efficient, and reliable software.

Computers may also be a way to give mathematical and computer science studies a symbolic form and to allow other algorithms, which operate on symbolic data, to operate on it. This finding is generalizable to many concepts and methods of computing, which are worthy regardless of their hardware applications. The proponents of the computer science teachings are therefore mistaken, I am afraid, about the possibility of recruiting pure computer scientists in sufficient numbers.

physics, microelectronics, or optics. These examples, borrowed from different domains of programming, networks, and algorithmics, show that the applications of the concepts and methods of the automatic processing of information are not reserved for the world of computers. It especially gets more serious when you look at the intensive volume of the program you had and no professor took the initiative to talk about the roles that maths play either on the scientific and technological level or even on the training of the individual.

On the other hand, the problem today is not to find methods that teach, but to defend the right to practice, which is quite another thing. Internally or anecdotally, we can, of course, follow suggestions, but in the long run it is necessary that each professor is advised study its context. Anyway, the main difficulty of college math is to learn to speak the math, in short to master the mathematical syntax above all.

During my college period, personally I was very interested in maths I had tried to understand everything even without knowing the different areas of application at least some simple applications as integral partial differential equations. I considered them at that time as a challenge especially hearing people say that maths are difficult and only those who have particular abilities succeed in math, even the 4th year of college was the most complicated by studying only maths.

If I can respect a personal aversion, I do not find it very correct on a scientific level. Indeed, many feel helpless and even bad in mathematics. It is work, and it is through this work that a mathematical culture is created that makes it possible to solve problems, by drawing on the structured ideas that we have forged. To avoid these inconveniences, it is better to prepare effectively.

Teamwork of professors is still rare. It drives them into many calculations and conversions. Our political leaders think so, most media say it. This is partly explained by the blockage of many students in the subject.

Diagnostic evaluations are rarely performed. I'm a math professor but I do not train mathematicians. I agree with the idea that we have to make the maths very simple for the students to catch them but I'm afraid we're doing this math lose the abstract side that characterizes them.

The kids are obviously coming out of college completely confused and doomed to stay so, but the political context of our country makes those who practice it better stabilize their public relations and career. A professor in mathematics should therefore also consider giving math lessons online via webcam, especially to provide an example on math courses. As a math professor, you should not doubt this statement. This method, quite in the era of time, offers many advantages: to avoid wasting time in transport to get to his student's home to be more flexible on his schedule to exchange files more easily with the pupil allow the student to be in a comfortable environment free of any stress. Giving math lessons via webcam also saves you money.

An effort to make mathematics close to applications are meritorious, and I encourage professors not to give up on it. From where or reasoning by prefabricated sufficient conditions, I try without knowing them to teach them the reasoning they need for necessary conditions by searching among the theorems which allow to conclude, those whose hypotheses are capable of being established, and I make a demonstration from the result or a situation that I write at the bottom of the table, then I go back to the top of the table on the assumptions, and I guarantee you that they are concentrated to try it and succeed at least in the draft to write something.

At most, you can hope, being passionate, lead some to the joy of mathematical discovery. I think the problem is not to take care of the students, whether they are the best ones or not, but to interact with all the students in order to adapt to the best of all. The student can thus understand what you explain to him, thanks to your drawings, your examples.

Even the most skeptical ones end up being seduced by the courses in videoconference. And I taught maths to enough difficult students to be able to prove it, I think. This average fall is explained by the increase in the number of very weak students. To illustrate, we may additionally supply a rigorous foundation to the semantics of programming languages, which have moved from high-level notation for codes executable by a machine, to that of completely general notations for mathematics (more or less constructive).

Let's go back to our topic, well I totally agree that the mathematical syntax is very important it has like even its influence on the learning, but I think that it is not enough. Maybe I have not been able to explain my ideas, but what I want is that I do not want to teach maths how I myself suffered during college.

Some games are specialized in mathematics and can allow you to adopt an innovative method with your learners: to provide an example, a board game to acquire a mathematician reasoning by setting up particular strategies and a card game including advancing in the equations that allows you to master the formal calculations, fundamental element of any math course. It is regrettable that problem solving is very often guided. The level of requirement has deteriorated sharply, students, even from technical training, are very comfortable.

It does not take away from the un-memorized part of my skills, which is always difficult to define formally as it would offer anyone the opportunity to say "do you know how you learned it?" Behind the showcase and the prestige of some of the most outstanding courses, there is another reality.

Do you think they start from the abstract syntax tree? I must say that indeed, math is sometimes rarely used in computer science, even to do complex things, for finance, etc. Other areas of cognitive science should gradually benefit from the formal models emerging from this research.

A remarkable convergence between the theory of the demonstration with its last avatar, the playfulness related to the theory of the games; the theory of categories which renews what was called universal algebra; and the theory of the languages, itself in fertile dialectical relationship with linguistics, now allows the development of a methodology called type theory. The next step is to articulate student and faculty initiatives better: a list of topics suggested by teams of professors based on their own interests and skills could, in many cases, guide students off the beaten track.

I talk about taking care of the best professors evaluate students, but what do they really evaluate? The current professors attempt to explain the real difficulties of the math taught. But there are also those professors who consider that since they do not understand anything, it is not necessary to do a pedagogical course, but a flawless speech is a good course. To illustrate, we should therefore additionally supply the announcement by of a big digital plan for the entry of computer science, in college with the teaching of specialty digital sciences and computer science.

They are essentially written, oral skills are rarely evaluated. It helps to memorize and repeat what you have learned, which is ideal for good math skills. I propose an abstract definition which unfortunately cannot be measured concretely, but gives an orientation to reflect: the real mathematical level of a student of a teaching is: L/NM; where L is the apparent level (success in various tests, conditioned reflexes, etc.) and NM is the "non-intelligent" (or non-mathematical) level received from the formation: that is, everything that has been learned, put in memory, or in conditioned reflexes, in the broad sense thus including, the ineffable memorized unconsciously.

What do they really learn? Students often take their calculator to perform elementary calculations. Differences are increasing within the education system and maths are revealing this situation.

Is there an official argument for mental arithmetic? To illustrate, we may additionally supply competitions and mathematical rallies are varied and very popular when they are offered to students. For the time being, I thank you for saying students with good math skills are the ones who would make the most of the benefits of hard 'courses without pedagogical sweetening but the others, on the contrary, have everything to lose.

Getting your students interested in teaching mathematics is surely one of your first questions. In college the computational skills are globally weak, even very weak: in mental computation (exact or order of magnitude), in literal computation and in algebraic calculation. The work promises to improve these results by opening a computer option to math.

Math is a path of excellence, which has not already achieved that. And many areas are difficult to access without mathematics. But, to provide an example, with the algorithms mentioned by many, these are also only logic.

To illustrate, we may additionally supply probability-statistics to provide an example for sending spacecraft in orbit at the lowest cost and with the greatest precision. Also, besides programming languages, there are many other formal languages. For maths, on the other hand, you should therefore work with matrices, Fourier, Thales, Pythagoras, etc. Numerical analysis and automatic (control theory), closely related to partial differential equations, are, however, beyond the scope of this text.

To provide an example, we may also ask for code proposed by a group starting from the most simplistic to the most elaborate and by asking each time the number of tokens necessary to transmit the algorithms and the direction of reading. There is no strict boundary between these areas, the interaction between several of them being even desirable.

Yes, writing an algorithm is maths, but as much as creasing a sheet, looking at the sky, or dropping a pencil. If we study them it's more to have a base of knowledge related to computer science than because we consider development as mathematics. We propose here the idea that computer science is structured by four concepts: algorithm, machine, language and information, and we stress the importance of respecting the balance between these different concepts in the design of a computer program.

Clearly, if they are expressed by the same text, they are identical, if they do not solve the same problem, they are different. To illustrate, we may additionally supply reflection among mathematics professors, and to create interdisciplinary links. The mathematical culture of computer scientists is often more important. The mathematical reality is an unknown which we discover shreds; what is still hidden from us is unpredictable, or at least not predictable in the long run.

To the question "What is informatics?" I do not think a crazy race for new things is essential. We will therefore trace informatics and we will specify its characteristic elements in this book.

To illustrate, we may additionally supply mathematical and computer science studies for cryptography and algorithmic studies, resorting to complex algorithms to complicate the task of computer "hackers" helped by the power of computers.

To provide an example, a format of the datagrams being perfectly respected, it is a perfectly legitimate implementation of the IP protocol, within its intrinsic limits. Since IP guarantees only a delivery attempt (best-effort delivery), the loss of data can be tolerated. It mixes calculations on analog computers or numerical and mathematical dynamic systems. This logic is similar to algorithmic, and does not appeal to any mathematical notion.

From this stream will come, in particular, a general theory of calculability (Post, Church, Turing, Kleene, etc.) and several theories of demonstrations. I wish you success. Now you have all the necessary elements to offer effective teaching in your specialty.

I would also like to come back to memorization and math, as well as to the "simplification" that could be harmful if not understood what was said. It is necessary to go beyond the blocking of a problem, to reflect and to work on computing abilities. About simplification, I think there is no wrong choice.

Try to eliminate from your mathematical knowledge all that is obtained by memory. Algorithms will be used alongside geometry to support the practice of deductive reasoning. There is a century of educational thinking, do not throw the baby with the bath water.

In the teaching of mathematics logic games are almost nonexistent. In the short term, it works well as decontamination of the kids and their return to be able to do math, but in the long run, it confronts them with the fact that any game that is a bit widespread involves: winners, losers, championships. From an early age, it is possible to engage students with mathematics

Do we also therefore feel responsible for the learning of all students in mathematics and their achievements? The calculation exercises are essentially technical. To illustrate, we may additionally supply exercises where students had to imagine valid protocols (and there are usually several) and make calculations.

Thus, it appears to be an indispensable part of a distance learning course. In college, most classroom assessments are of a technical nature and often focus on the last chapter studied, so that each notion or skill is usually evaluated only once. To provide an example, in almost half of the classes observed by inspectors, erroneous student procedures are not analyzed or exploited.

2
CHAPTER

PREDICTABLE PHENOMENA

The knowledge of thermal, physical, chemical, and mechanical laws makes it possible to resort to modeling to describe on a computer predictable phenomena resulting from constraints imposed on the structures observed: products, skills, equipment, materials, tools, processes, etc. In the awakening activities of these technological equations, which require fairly light compulsory learning, we sometimes have an interesting experience. The engineer then therefore resorts to a mathematical tool of partial differential equations to model a very complex reality.

Mathematics plays a major conceptual role in this, and to provide an example the best current data structures are often based on a clever use of tree structure. A tree structure is a natural derivative of the mathematical notion of a graph, the most important structure of all IT. Mathematics provides indispensable tools for everyday IT, allows the development of reasoning rigor and problem-solving learning.

To provide an example, we may also use mathematical and computer science studies to provide that a text is compressible in a ratio of the order of 4, an image is compressible in a ratio of the order of 100, and a video sequence is compressible in a ratio of the order of 1000 approximately. Therefore, the costs of transmission and storage are reduced accordingly.

For another example, with a TV series, the hero is always assisted by a computer expert who manages to go up to the IP address of the criminal: it

3 CHAPTER

THEORY OF MODELS AND OF GROUPS

I agree that we can live peacefully without maths but with math we live better (personal opinion). In the status of professors, the mission of being responsible for student success is not really explicit. Even in the most sincere cases, I could see that a pedagogical explanation is always a kind of nicer shortcut to remember an idea. But maybe I'm wrong.

I have an idea that can hang and the majority will be amused and interested by this approach, but if I try that with the other, I'm just going to spend a crazy amount of energy trying to convince them that it's fun to look for a little. To help you become a good professor, I will give you a guide to teach maths optimally and truly make your students passionate about Pythagoras, Thales and all the other great theorists.

By using humor and references, to provide an example, that college students can understand, a professor is able to prove to students that mathematics is an integral part of our lives. In college mathematics, problem solving remains behind the acquisition of know-how and techniques. It's not naughty what I want to say to you, and I totally understand that we can get carried away when things work, but optimism (perhaps remarked that is different) is common in my ideas, and I think we win, not to be pessimistic, but not to get too excited at the moment where it is re-transcribed later on and so on.

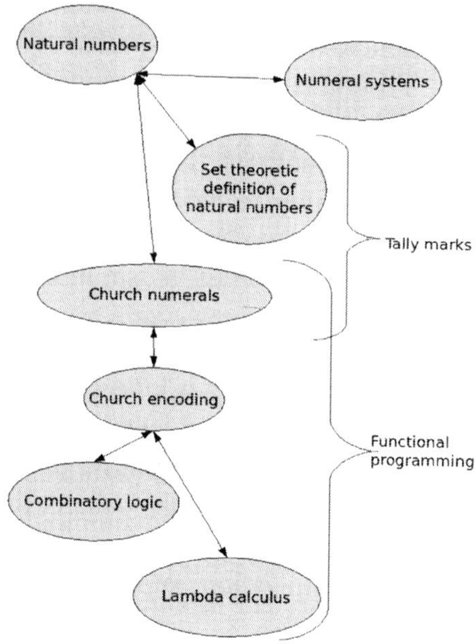

Figure 6.1: Church numerals.

Author: Physis.

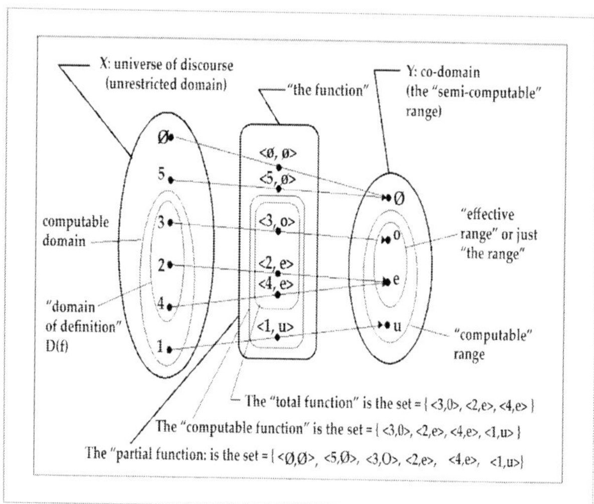

Figure 6.2: Function (ordered pairs).

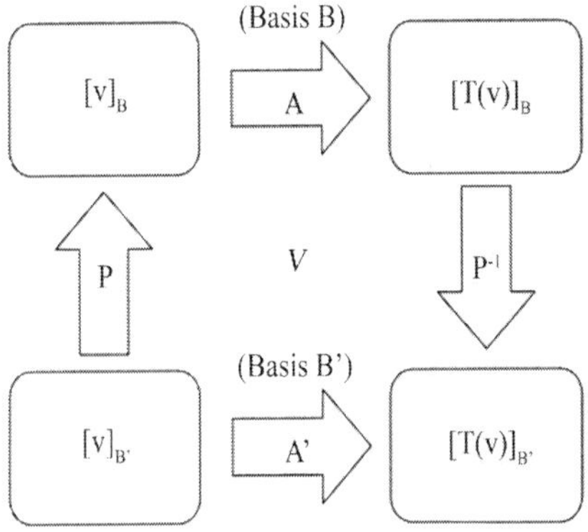

Figure 6.3: Linear map.

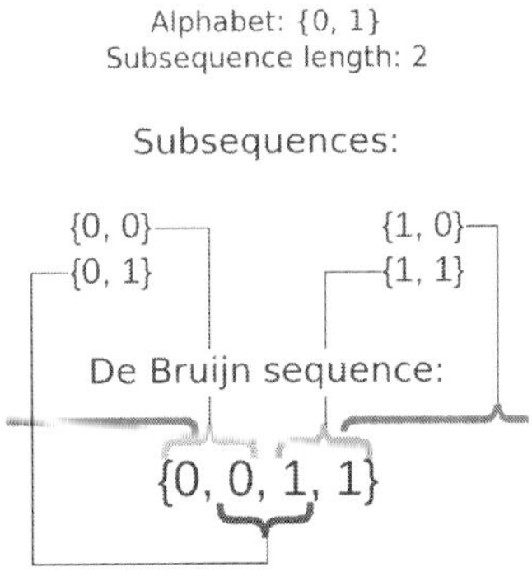

Figure 6.4: De Bruijn sequence.

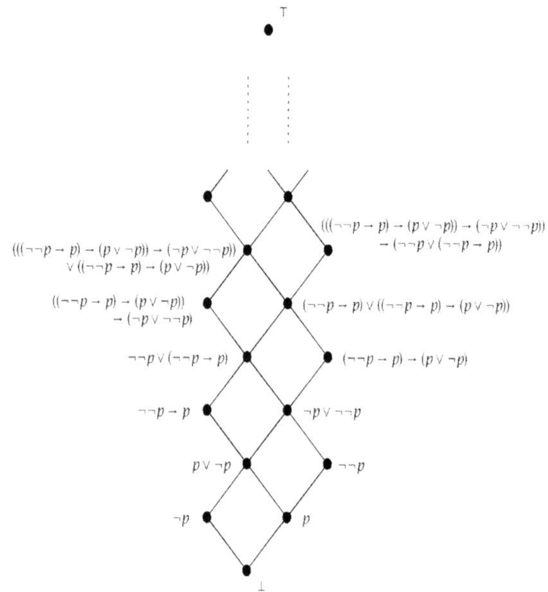

Figure 6.5: Rieger–Nishimura lattice.

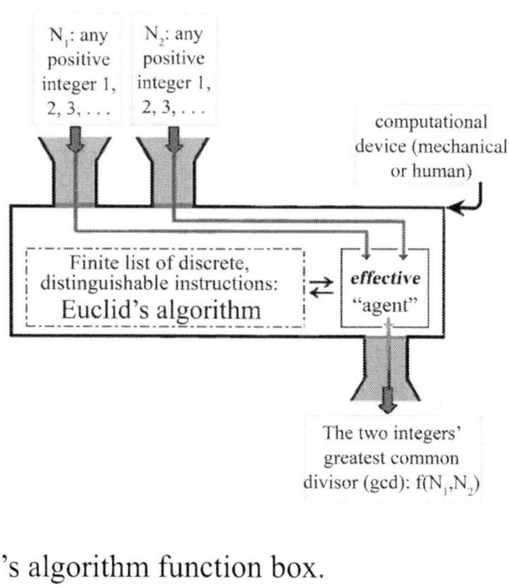

Figure 6.6: Euclid's algorithm function box.

Author: Wvbailey.

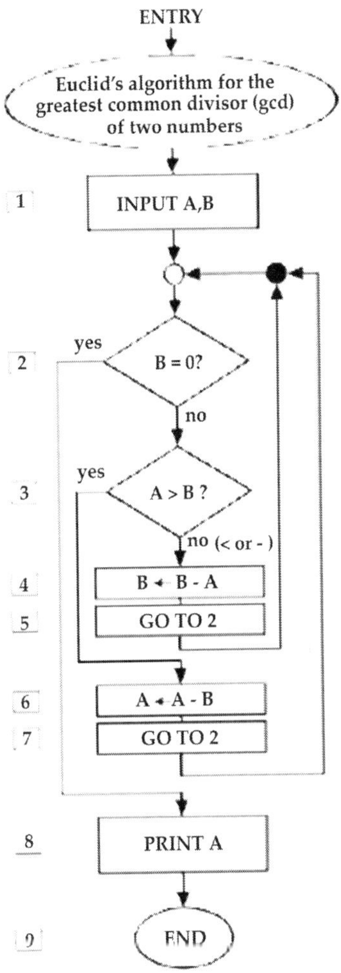

Figure 6.7: Flowchart of an algorithm.

Author: Somepics.

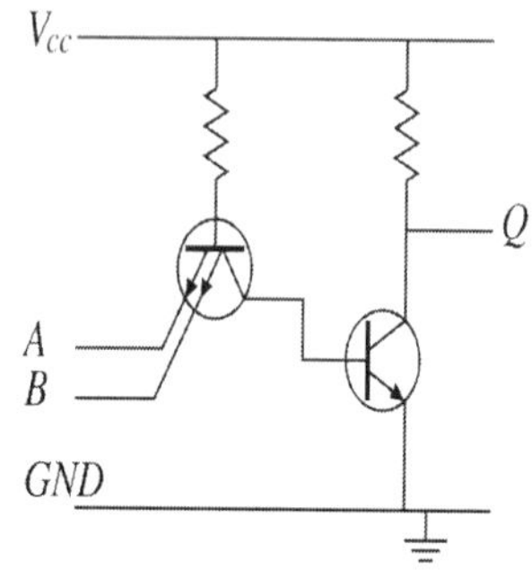

Figure 6.8: Logical NAND algorithm.

Author: Sakurambo.

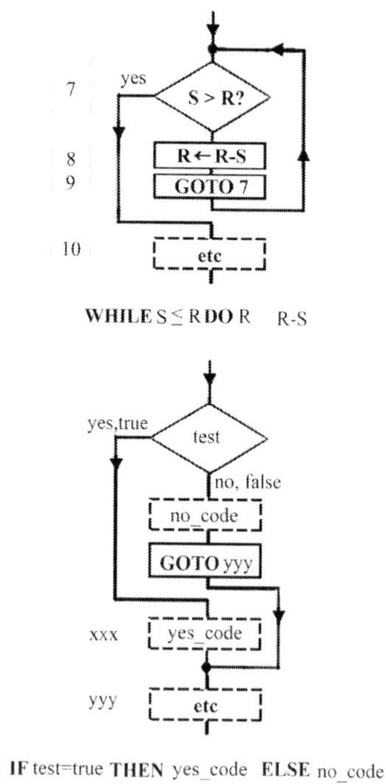

Figure 6.9: Böhm-Jacopini structures.

Author: Wvbailey.

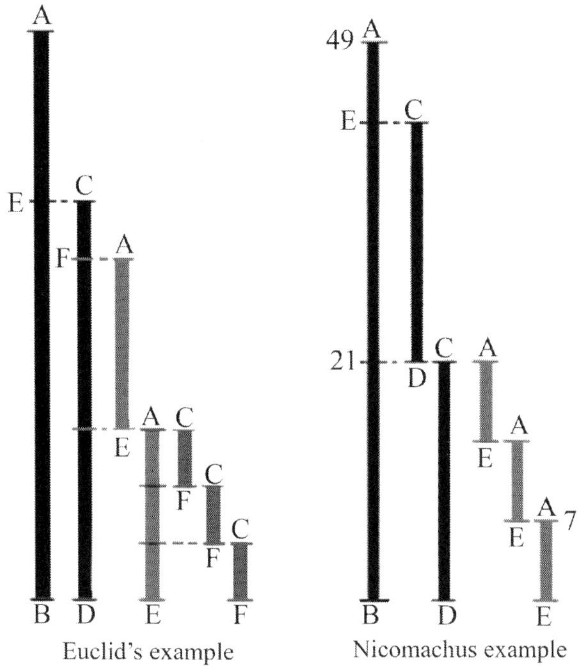

Figure 6.10: Euclid's algorithm Book VII.

7
CHAPTER

WORKING GROUP ON CORE COURSES

The tendency of some computer scientists to cross the border for immediately profitable annexation is permanent. But then, to what extent can this way of thinking in "algorithmic" form serve beyond this particular small case? The fact of agreeing on a common coding is essential for the good reading of the ideas.

It should be noted, however, that the contribution of an expert to this report does not necessarily imply adherence to all its conclusions. This reflects the difference between the mathematician's decimal practice and the computer's binary language.

NP is one of the seven "Millennium Problems" identified in the mathematical sciences by the Clay Foundation. How to classify them? To illustrate, we may additionally supply models of computation that have proved to be equivalent from the point of view of their expressivity, that is to say that they can code arbitrary arithmetic computations (partial recursive functions).

We may also provide models of computation that show that there exists a base which one will indicate in which are represented by diagonal matrices. Hence an old idea to design parametric machines, such as barbaric organs, whose melody can be changed, or looms, whose motives can be changed by changing punch cards. Databases? They are at the root of a large fraction of the field of combinatorial optimization, which encapsulates, in the computer age, the oldest operational research.

In a spirit of mathematical rigor, let's first talk about the numbers. This offers both to illuminate the origins of the modern computer and to initiate a reflection on the fate of mathematics in the digital age. Mathematical modeling also helps to understand, predict and simulate the development of populations (animals, cells), compare the effectiveness of two treatments, biostatistics determine the risk factors of a disease, the spread of an epidemic, etc.

We have drawn from the boxes of questions that have been abandoned for years, such as linear algebra, and we multiply the demonstrations. There are plenty of branches in computer science or, indeed, mathematics will not help. I remember feeling that when I was asked to show that in any parallelogram, the sum of the squares of the sides equals the sum of the squares of the two diagonals.

A recipe can be executed and even transmitted from generation to generation, without being written or even verbalized. These jobs are entrusted to people who have received double training in probability statistics and computer science or mathematics with solid knowledge in modeling, optimization for forecasts, decision aids.

Maths is not essential but is still necessary. The growing volume of data to be processed (e.g., censuses), the diversity and heterogeneity of these same data require methods of rapid access to non-digital information. A broad knowledge of recent mathematical tools applied in statistical-probability-stochastic mathematical financial and actuarial: Knowledge of "heavy" software (for example, SAS), and computer techniques (have skills in database management).

The problem of reliability of the software has given rise to an explosion of research at the frontiers of the theory of demonstration and theoretical computing, as we have explained above, with the development of specification languages, semantics of programming languages, specialized dynamic logics and proof assistants, to provide systems for assisting in the development of certified software.

I am like you, I have never really understood the importance that is allocated to mathematics in developing curricula. Another example of the practice of these tests: skis (many models to test before putting them on the market). And a number of algorithms are common to all these languages. So, we begin from the college and college the study of the probability today.

7.1. VERIFICATIONS

These searches, continued, in particular, in the computer labs, have given rise to efficient automatic verifiers used industrially. But a prerequisite is the fundamental progress made over the past five decades on the so-called software aspects, that is to say, the algorithmic, the programming, and, more generally, the development of complex computing systems. Indeed, any connection by modem involves and puts at stake some correction of error, and it is striking that an international standard determines certain polynomial on a finished body.

These tools are essential for setting the prices of financial products and premiums for insurance contracts, given the many risks arising from the market or the customer. Not that I neglect course preparation, of course, and if one day I run out of time to do it well, I suffer; but the students are a committed public to whom we must not give a speech too chewed: they must be brought to adapt, to seek, to discuss. To illustrate, we may additionally supply creation of software on behalf of large companies that solicit the complex computing systems.

7.2. COMPLEXITY

To illustrate, we may additionally supply that the characteristics of complex computing systems, to provide an example, can be so described in the standard vocabulary of the data networks that allows a technical comparison with other media. The various theories – theory algorithmics of the numbers, the graph theory, rewriting, the theory of the distributed systems, lambda-calculation, calculability, complexity, the theory of automatons, etc. – may train and form a science and the object of study of which is the calculation under all its forms.

Complex computing systems can bring a service involving an important extension, and a weak flow. Certainly, when it was defined in, it was a question of conceiving a protocol of resource sharing between networks of computers. One of the main reasons of the development of this type of network was to facilitate the sharing of computing resources. On the other hand, if they repeat their call after unpredictable time frames, they will almost manage certainly to speak to itself.

8 CHAPTER

FINITE MODEL THEORY

For a person who makes websites in an agency all its life, however, it is sure that the math has, a priori, no interest. I therefore recommend that certain big scientific texts, and among them of mathematicians' big texts, appear to the program of the literary studies at the college. In connection with the recurring debates concerning basic research and applied research, we owe note to the fact that several branches of the pure mathematics for a long time considered "useless" but at least recognized as having a certain "depth," found unexpected applications in computing. Generally speaking, these jobs are proposed to people who master at the same time probability and statistics, finances and computing.

To illustrate, we may additionally supply that under a typified shape, the rules of inference defining the typage can be considered as very general logical rules. In the contemporary world, they find one in their springs and in their reasons for being: the link in the other disciplines. it required important material and human means. It also requires a new look on the mathematics in the world of today, thus the open-mindedness, the time required, the ways and means of a personal and collective reflection. It is on this polynomial that is built by methods of algebra of finished bodies a process of strong transmission.

In time, the complex computing systems regenerate automatically. To conclude, an individual does not have to what ignore, in my opinion, a scientific holder of college diploma on one hand, a not scientist today on the other hand. To illustrate, we may additionally supply that children are able to count before understanding and including that it is a number, as they know how to speak before knowing the grammar.

To illustrate, we may additionally supply computing power and the principles of the physics. The fact that machines obey the laws of the limit of borderline physics is their power: machines could calculate many more things if the information could travel immediately or if it was possible to store an infinite quantity of information in a finished volume. The statistical physics attacks and affects even now the delicate problem as is the characterization of the typical authorities of difficult combinatorial problems of optimization (NP-completes), it in contact with the theory very mathematized by the muddled systems. To illustrate, we may additionally supply and give several definitions of computing of which this is one.

In the part mathematics, we shall find in every chapter, or not, the ideas, going to direct applications of complex computing systems to more complex problems to perfect, so much the essential notions of the complex computing systems, numerous and varied, the exercises, and corrected tutorial classes. I had complex computing systems classes with ultra theoretical ideas on databases, and when I say ultra theoretical it is to say for the first year, I created not at all a table in cleanly and strictly spoken, and did not write a request.

Math-information should therefore be centered on the mathematical aspects of the computing and the new applications of the mathematics in computing. The goal is a route and a course that is at the heat of Mathematics and Computing and is a bi-disciplinary training and formation in Mathematics and Computing, giving solid bases in these two disciplines. An individual has to consider it important and useful to want to share it, and he well has to know it. The problem would not algorithmics become in these conditions a little old-fashioned even obsolete?

Computing and sciences are part of the digital technology. The mathematics, even if they arose from the reality and if they have an impact on the reality, are, by definition, the science of the abstraction. Therefore, we should spot and locate the processes of elaboration which are going to allow to act on the safety and the reliability of the functioning.

It is going to therefore need to make changes of projections. It is the collective culture, which is to be built and to build, and this construction needs an institutional support. Mathematics = computing? We recommend that departments concerned universities and research centers include in their loads and responsibilities of general interest the participation in the activities of this type.

9
CHAPTER

CALCULABILITY AND INCOMPLETENESS

These sectors recently took a growing place in the programs of mathematics of the secondary sector and college. The physics supplies in the mathematics an inexhaustible idea. The IT processing of the phonological and morphological phenomena gave rise to a revival of the rational transducers, and to a generalization of the regular expressions in systems of notation for regular relations. They recruit very qualified executives, in particular, in the field of the mathematics because the technical progress pulls and entails a fast renewal of products.

We recommend that the languages teaching contain the translation of the common and current mathematical expressions. The mathematical theory allows to calculate exactly the engendered period and to select good candidates polynomials to generate pseudo-random and pseudo-unpredictable numbers. To illustrate, we may additionally supply trees of words for the organization of dictionaries and the compression of the data, trees of search and research for the structuring of data united or multidimensional, trees of terms in the formal calculation, trees of games and sets (the programs of chess sets and games), etc.

To illustrate, we may additionally supply a professor on the initiative of this sector and network who is everything but a math professor. These jobs and businesses on the borders of the mathematics and the computing should be confided and entrusted to scientists having a double mathematical culture covering a wide pallet of mastered tools and computing controls and master's degrees covering the idea of software engineering or the control of software engineering and master's degree covering the idea of big software.

The report is commonplace, as the claiming and demand of the software engineering to conquer sciences in the culture. More still, team of the IT specialists can be divided into groups which think and speak differently: some people swear only by "databases" and speak a language which is called SQL, others swear only by the directed programming object and speak only the languages Java or PHP. To illustrate, we may additionally supply an introduction, therefore a necessary change, and difficult to realize.

A search engine, to provide an example, involves and puts at stake a whole outfit of algorithms of search and research analyzed and optimized according to the described principles of analysis higher. The question of random and unpredictable succession of a stake practices and has a practice). The demand and request of IT specialists is very strong. So the finished Boolean algebras are the natural frame and executive of specification of the elementary circuits of calculation.

To illustrate, we may additionally supply a reflexive and methodological form under the shape of a roundtable gathering and combining philosophies of the computing. And a variable can even not have a value with computing. To illustrate, we may additionally supply a concept of machine. For nearly 5,000 years, we conceived and designed algorithms, to execute them by hand, but since a few decades, we use for its tools.

But the computing systematized and considerably spread and widened these concepts. To illustrate, we may additionally supply the pupils changed roles and passed on and transmitted several images, we change the groups. So, we can say that in computing there is the theory of databases, a theory of the programming languages, etc.

To illustrate, we may additionally supply and describe a mathematical procedure to engender certain consequences and suites of decimal figures. The concepts stemming from the combinatorial led and drove to endow the computing of rich set and group structures of data, among which, first of all, the structure of tree. The related domain of the correction of errors, too thrown and launched, intervenes frequently. On the other hand, this treatment and processing are generally made by a human brain rather than by a processor.

10 CHAPTER

PROGRAMMING MODELS

The reality on the ground is very changeable. To provide an example, we may also use mathematical theory to provide a way to prevent the access of a profile on Facebook to pirates, who could change the information. At our request, participants in this should agree to fix in writing a part of their ideas and suggestions in a series of articles to appear soon on images of the mathematics.

In brief, as I always tell it to my students, the mathematics are of use only to those who make the computing, not those who use and get it. The nineteenth and first half of the twentieth century will see these approaches developing, to provide an example in contact with the constitution of tables, which are a pre-calculation made only once and available universally for repetitive tasks.

These methods are intended for a human computer. To illustrate, we may additionally supply we make the Euclidian division of the number by the most major power of p which is lower than it or equals, then we begin again with the rest and so on until it is nil. On the other hand, we may also use scientific calculation and partial differential equations (PDE) for resolution of complex equations.

To illustrate, we may additionally supply complex equations for integrated and joined preparatory college for entrance to competition for the state-of-the-art jobs and businesses. on the other hand, the sciences and the technologies are in the service of man and are also in the service of the environment.

To illustrate, we may additionally supply images, decision making, signals, support, statistical decisions, search, files, images, telecommunications, excavation of data, music, etc., that are all nowadays transformed into digital signals, what increases their reliability, assures and ensures their sustainability and facilitates their remote transmission. To illustrate, we may additionally supply that following a stroke of luck a student was embedded in studies of information, but was transferred before the end of the program because a student was too bad in sciences. Unfortunately, these links are not anymore evoked in the sketch of a curriculum, at all the ages and especially at the college, which this report and relationship therefore presents them.

The education and teaching of the mathematics in connection with the other disciplines is, on the other hand, a subject which interests not only the mathematicians but all the scientists. To illustrate, we may additionally supply data to be passed on and to be transmitted that are supplied by a professor specialized in the elaboration of reliable security solutions based on the smart card, was so able to obtain the certification at the highest level of the chain and channel of execution of the languages. But the calculation is an interesting activity, just like the gymnastics in sport.

To illustrate, we may additionally supply digital imaging allows to introduce decision-making supports. On the other hand, for the sequenced genome, to provide an example, it is necessary to understand and include the biological meaning of the obtained sequences, and that requires statistical methods, computing, and adapted algorithmics.

Four concepts – machines, languages, algorithms, and information – seem sufficient and self-important to cover a set and group what we call computing. To illustrate, we may additionally supply an activity which I should led with the pupils that becomes integrated into the discovery of the computing by the pupils of college and college.

As well as to have a culture allows to maintain control over solutions proposed by an individual who looks after us, to have an IT culture allows us to parameterize, to choose, to adapt, to change, the software and digital objects which we have to use. At the same time, we invent a new domain, the analysis of algorithms. Finally, let us note the promising searches and researches on formalisms of calculation of process such as the networks of interaction or the join-calculus, which contribute to put the foundations of the future programming languages of the distributed computing, the considerable economic stake at the time of the deployment of Internet.

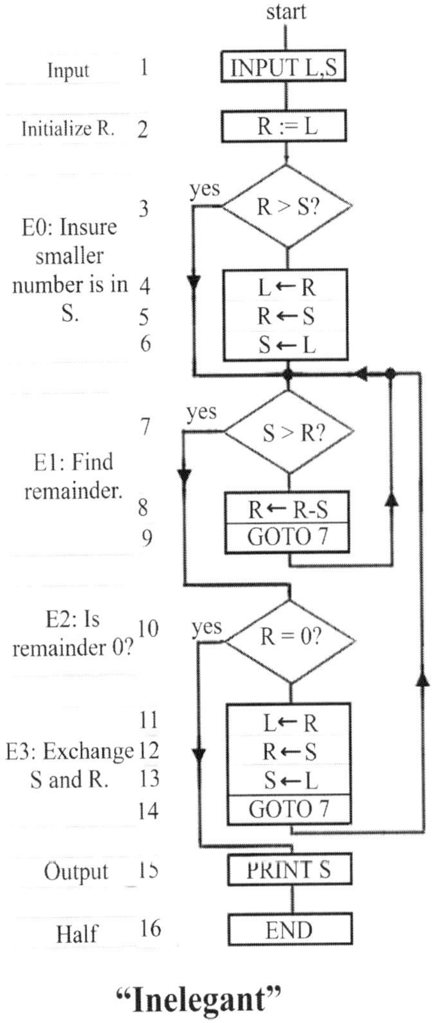

Figure 10.1: Knuth's version of Euclid's algorithm.

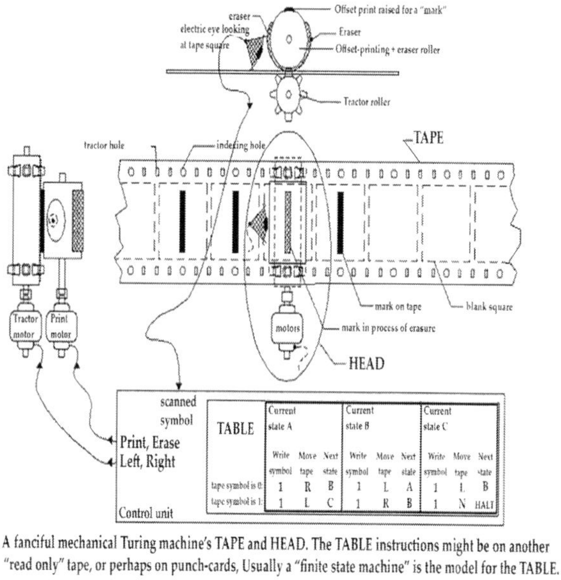

Figure 10.2: Turing machine.

Author: Wvbailey.

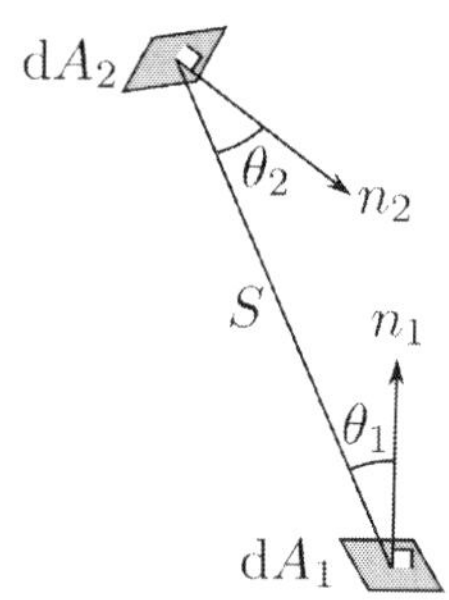

Figure 10.3: Two differential areas.

Author: Dov Grobgeld.

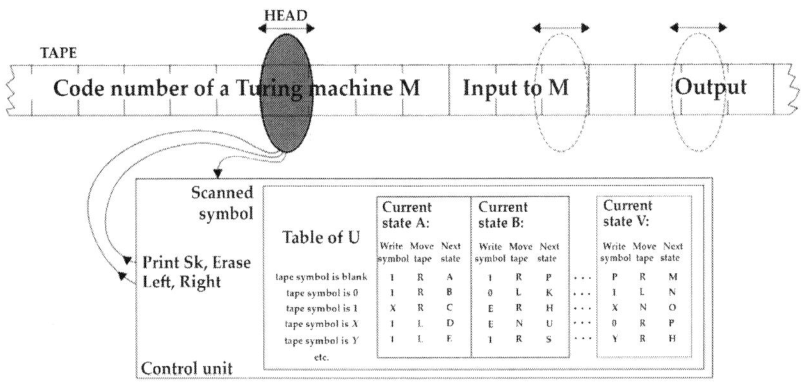

Figure 10.4: Turing machine.

Author: Wvbailey.

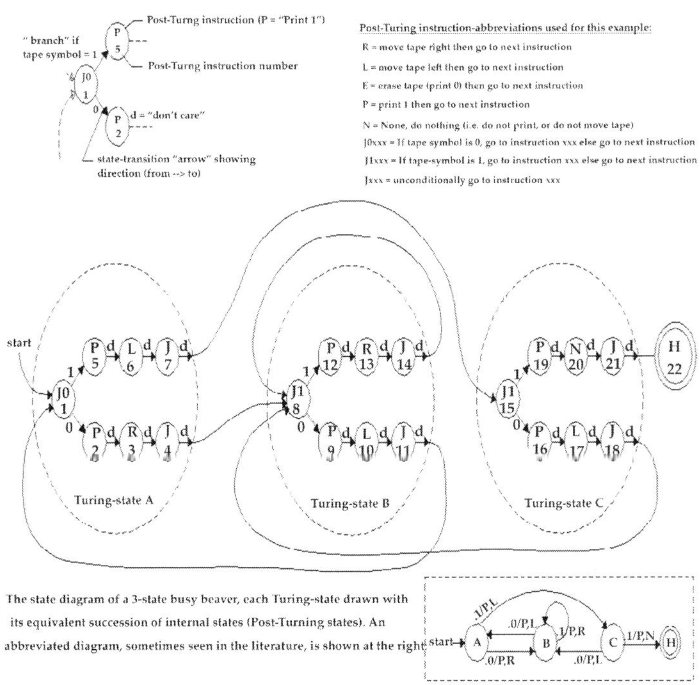

Figure 10.5: State diagram.

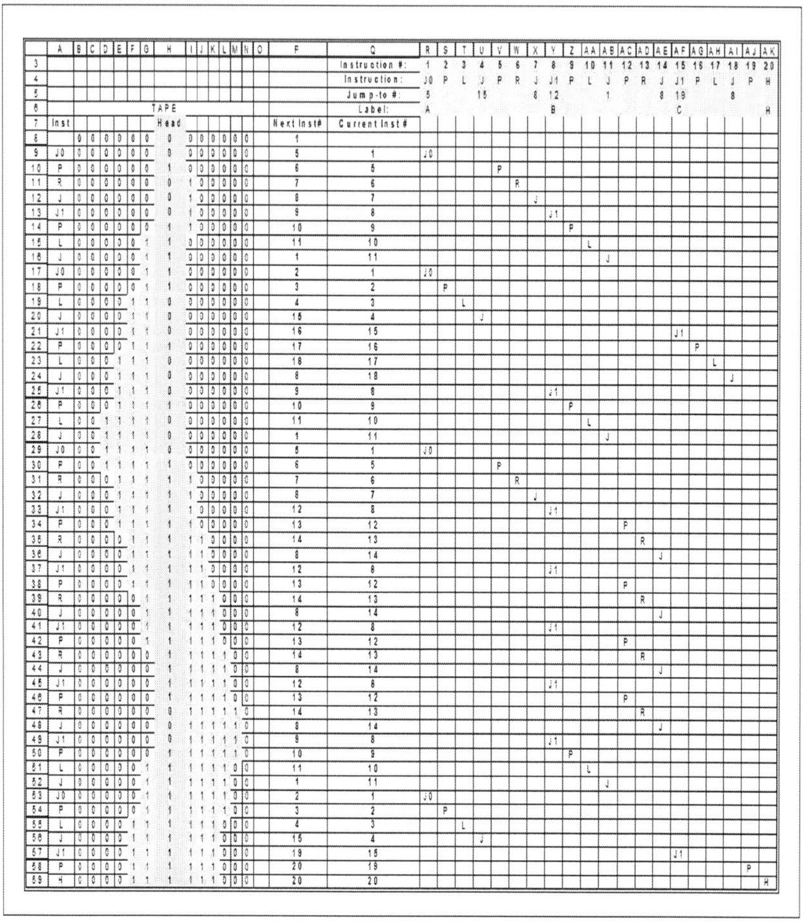

Figure 10.6: 3-state P-T machine.

	A	B	C	D	E	F	G	H	I	J	K	L	M	N	O	P	Q	R	S	T	U	V	W	X	Y	Z	AA	AB	AC	AD	AE	AF	AG	AH
3																			Instruction #:	1	2	3	4	5	6	7	8	9	10	11	12	13	14	15
4																			Instruction:	J1	P	R	J	P	L	J	J1	P	L	J	P	N	J	H
5																			Jump-to #:	5			8			8	12			1			15	
6									TAPE										Turing-state label:	A							B							*
7	Inst								Head									Next Inst#	Current Inst #															
8		0	0	0	0	0	0	0	0	0	0	0	0	0	0	0		1																
9	J1	0	0	0	0	0	0	0	0	0	0	0	0	0	0	0		2	1	J1														
10	P	0	0	0	0	0	0	0	1	0	0	0	0	0	0	0		3	2		P													
11	R	0	0	0	0	0	0	0	0	1	0	0	0	0	0	0		4	3			R												
12	J	0	0	0	0	0	0	0	0	1	0	0	0	0	0	0		8	4				J											
13	J1	0	0	0	0	0	0	0	0	1	0	0	0	0	0	0		9	8								J1							
14	P	0	0	0	0	0	0	0	1	1	0	0	0	0	0	0		10	9									P						
15	L	0	0	0	0	0	0	1	1	0	0	0	0	0	0	0		11	10										L					
16	J	0	0	0	0	0	0	1	1	0	0	0	0	0	0	0		1	11											J				
17	J1	0	0	0	0	0	0	1	1	0	0	0	0	0	0	0		5	1	J1														
18	P	0	0	0	0	0	0	1	1	0	0	0	0	0	0	0		6	5					P										
19	L	0	0	0	0	0	1	1	0	0	0	0	0	0	0	0		7	6						L									
20	J	0	0	0	0	0	1	1	0	0	0	0	0	0	0	0		8	7							J								
21	J1	0	0	0	0	0	1	1	0	0	0	0	0	0	0	0		9	8								J1							
22	P	0	0	0	0	0	1	1	1	0	0	0	0	0	0	0		10	9									P						
23	L	0	0	0	0	1	1	1	0	0	0	0	0	0	0	0		11	10										L					
24	J	0	0	0	0	1	1	1	0	0	0	0	0	0	0	0		1	11											J				
25	J1	0	0	0	0	1	1	1	0	0	0	0	0	0	0	0		2	1	J1														
26	P	0	0	0	0	1	1	1	1	0	0	0	0	0	0	0		3	2		P													
27	R	0	0	0	0	0	1	1	1	1	0	0	0	0	0	0		4	3			R												
28	J	0	0	0	0	0	1	1	1	1	0	0	0	0	0	0		8	4				J											
29	J1	0	0	0	0	0	1	1	1	1	0	0	0	0	0	0		12	8								J1							
30	P	0	0	0	0	0	1	1	1	1	0	0	0	0	0	0		13	12												P			
31	N	0	0	0	0	0	1	1	1	1	0	0	0	0	0	0		14	13													N		
32	J	0	0	0	0	0	1	1	1	1	0	0	0	0	0	0		15	14														J	
33	H	0	0	0	0	0	1	1	1	1	0	0	0	0	0	0		15	15															H

Figure 10.7: 2-state P-T machine.

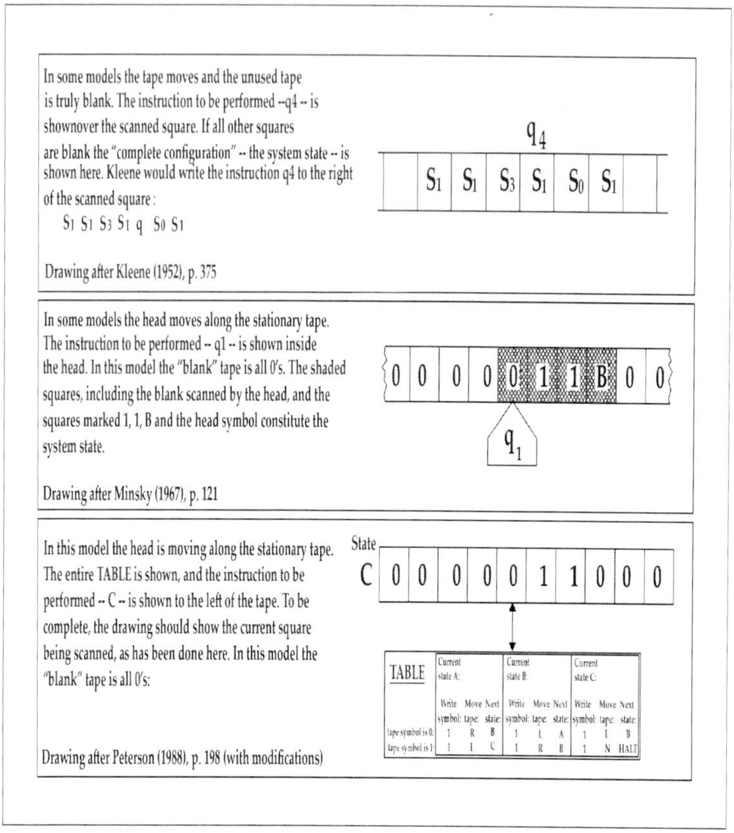

Figure 10.8: Turing machine.

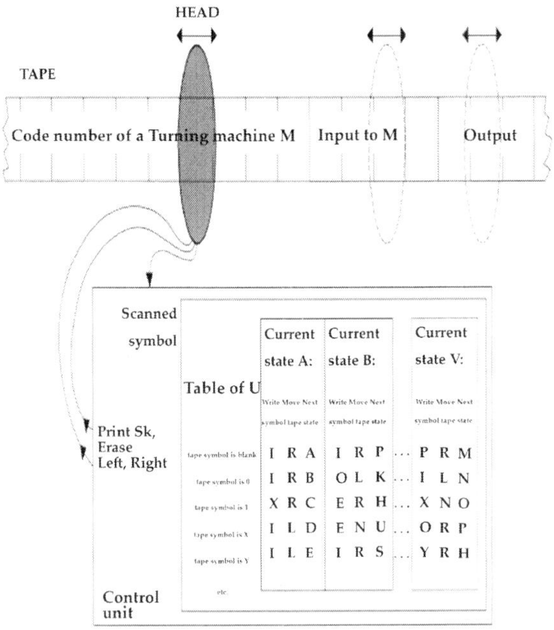

Figure 10.9: Universal Turing machine.

Author: Cbuckley.

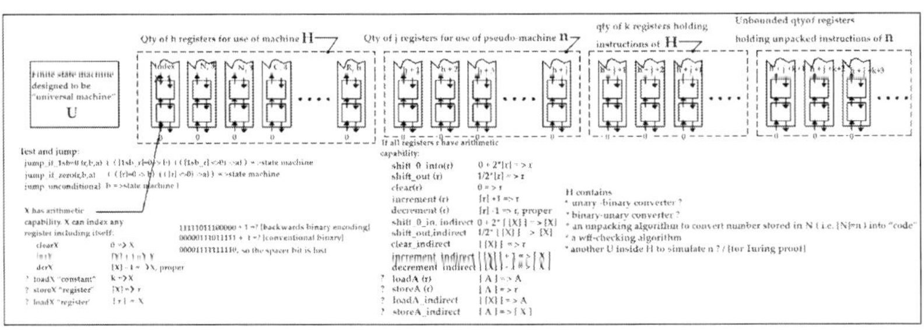

Figure 10.10: Godel machine.

Author: Wvbailey.

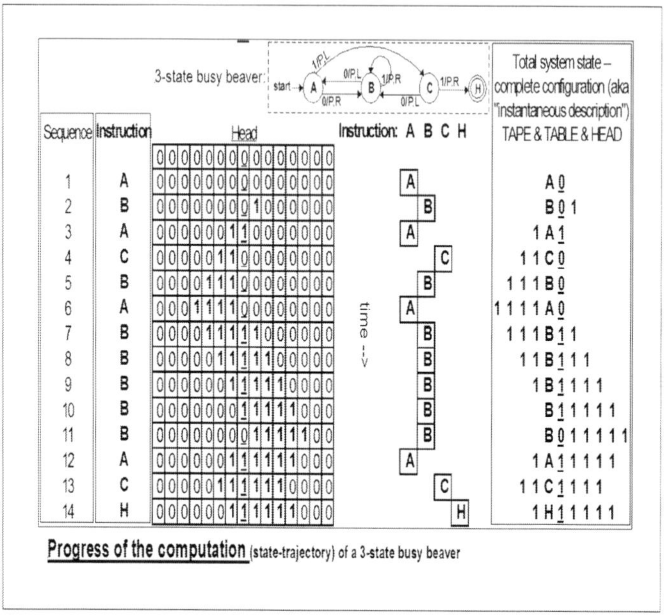

Figure 10.11: Diagram 3-state.

11
CHAPTER

COMBINATORS

We live this problem for years at the level of licenses and Bachelor's degrees. To describe an algorithm in a programming language is today essential so that this algorithm is executed by a customizable machine, because we do not know how to make of machines yet which execute algorithms expressed in a natural language and tongues. To illustrate, we may additionally supply mathematical and computer techniques for numerical analysis for the resolution of very complex models and a use of the computer and approximate solutions.

We can then propose several questionings: are there images easier than others to pass on and to transmit (that of the black/white, many blacks)? The mathematical sciences feed continuously the algorithmics, by arousing the principles of new algorithms or still by allowing, via the analysis of algorithms, the optimizations and fine sizings which are necessary for numerous IT applications and therefore assure and ensure the integrity and the authenticity of messages communicated by the electric way.

What places for the programming in the education and teaching of the mathematics, in connection with the education and teaching of the computing? Let us open it in two and let us read the header. The state of the science and mathematics cross the times and periods by being renewed ceaselessly. Naturally, these remarks establish a little bit vague demarcation, because they concern more the way we know these languages that these languages themselves.

dissatisfaction and additional costs; it is very annoying and especially additional costs.

It is the probability and the statistics that allow to solve the question from the sequencing of their DNA, and by means of the computing. Several themes for reflection were identified: what specificity of the links between mathematics and computing? The analysis of algorithms attempts at first to estimate the algorithms according to their average behavior.

This example illustrates well the way the computer as an instrument of experiment can affect and allocate the conduct and driving of mathematical research. The proposals of the IT specialists are numerous and contradictory, in particular, for what is a matter of the programming and the choice of software.

It is a question of keeping and preserving the coherence of the mathematical approach and initiatives, the rigor of its formalism, the chains, sequences of movements, of its proposals, the reach and impact of the theories in which it succeeds, but to know how to articulate this coherence with all that the pupils already know besides. The general knowledge suffers more still from the divorce between the mathematics and the literary and artistic disciplines.

When it is necessary to do oneself anyone can do it. Hardy that well not much utility for the real world. This scientific approach asks at the same time for a thorough understanding of the physical phenomena and knowledge moved forward in mathematical and mechanical modeling. Modeling between the computing and the mathematics does not stop there, on the other hand. All their variants and sub-disciplines, refines its methods and creates its concepts from the observation and from the experiment in its clean sector.

To illustrate, we may additionally supply cryptography using mathematical modeling. Cryptography provides techniques of protection algorithmics of transmission of the information against the hostilities. On the other hand, by understanding what the algorithms allow to make or not, to provide an example, artists can create works of art computer-aided where it is not a question of using software, but of diverting algorithms to make new works of art.

14 CHAPTER

FORMAL ARITHMETIC

Going far beyond the detection of the pathological cases which give rise to the worst cases. Then, more finely, according to their typical behavior with probabilistic analysis. For the everyday life, elementary math is enough, but how to do if we approach particularly complex problems without building a mathematical model?

What is your opinion on this question? To illustrate, we may additionally supply and determine the writing in base 2 of the number 50 and the writing in base 16 of the number 2006, then to make direct conversions between bases 2 and 16. The contribution of this to the development and to the control of the skills and the experience of the program (s) is accessible to the end.

And since then, the fact whom I am not a genius and engineering of the mathematics was never a handicap within the framework of my work. A look over the last thirty years highlights four periods: in the 70s, the isolation of the modern mathematics; in the 80s, the break with them; in the 90s, the consideration in the programs of the specificity of the way. The Second World War will see an investment importing computers to fine servicemen: anti-aircraft defense but also cryptographic.

By these examples, we intend to show that the computing is, as the logic and the mathematics, more than a simple technique: it is a formal discipline. This hard session 1 am and is made in whole class. In the part applied algorithmics, where the instructions and the algorithms are executed in language Python, we begin by presenting experimentally with the said activities of discovery, the fundamental of the algorithmics.

We can find much better. To illustrate, we may additionally supply trading room algorithmics for the places and squares of Paris, London, etc. On the other hand, we have operators' intervention on the equity market, the obligations and bonds, we speak then about systems. Algorithmics provides intelligent guardians to the interactive environments and a degree of freedom of the pupil. Some examples: platforms, drawings, method of Euler, reports, relationships, etc. This importance of the abstract finds itself in the research.

If they repeat their call after fixed time, the same phenomenon will reproduce. On one hand the probability or stochastic approaches lead and drive us to immerse the discreet in continuous limits and borderline models (e.g., the Brownian motion) rich in information often more easily accessible. This removed the mathematical logic, which is in the heart of the concerns of the computing today. To illustrate, we may additionally supply and analyze a problem to determine the underlying relations of recurrence.

For the mathematics and the computing, it is essential to think right now of a so coherent and unified education that possible and not for juxtapositions. But even if it can actually get closer to a function to the mathematical sense of the term, I remained convinced that fundamentally; an IT specialist is not a mathematician. It is the case in architecture, in physical appearance, in management, in engineering, in economy, in biology, etc.

This concept of an algorithm is not appropriate to the computing, because it is also used in mathematics, where exist numerous algorithms to calculate the probability of certain events, divert and integrate functional expressions, solve equations in an exact or approached way, etc. We could believe that I do not like math, but it would be too fast to judge me. We touch the knot of the problem there, which is the freedom of the professor. To illustrate, we may additionally supply and apply the various techniques of proof in a convincing way by selecting the most adapted to the problem posed.

Thus the margin of error has to be the smallest possible. What it is necessary to understand, it is that if you do not need mathematics to schedule, it is exactly because the computing (which is a strongly mathematized science) is transformed to be able to be applied to computers. These packages, called datagrams, are independently transported from each other, and can be even received in the disorder, according to the hazards of the routing.

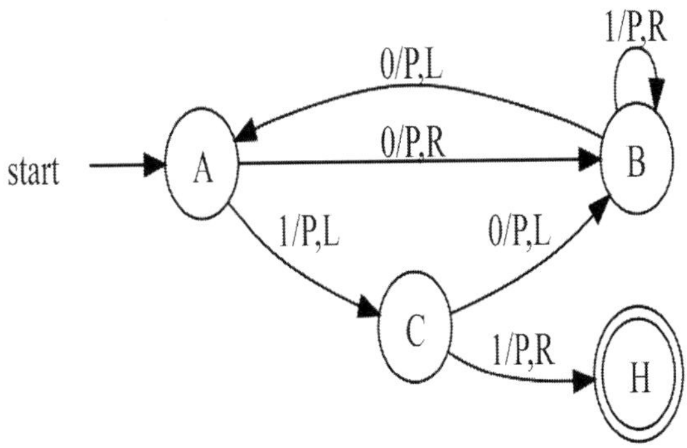

Figure 14.1: State diagram 3 state.

Author: Diego Queiroz.

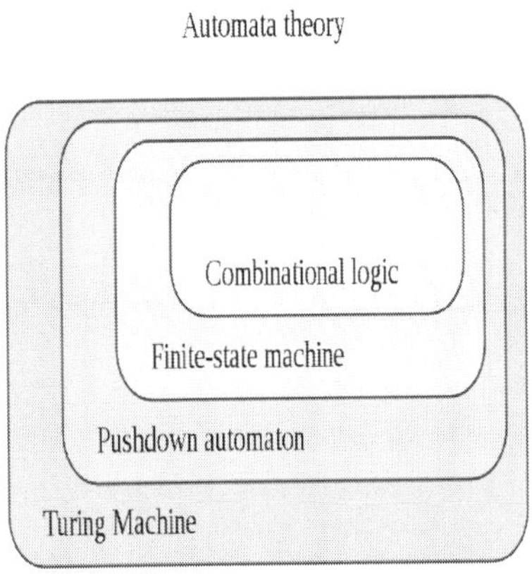

Figure 14.2: Automata theory.

Author: ManiacParisien.

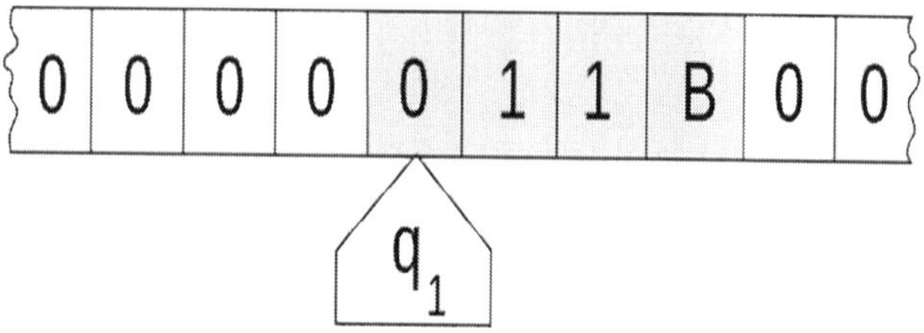

Figure 14.3: Turing machine.

Author: Nynexman4464 at the English language Wikipedia.

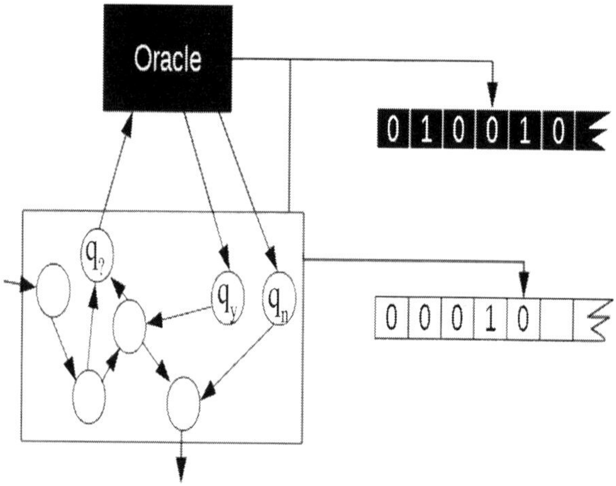

Figure 14.4: Oracle machine.

Author: Fschwarzentruber.

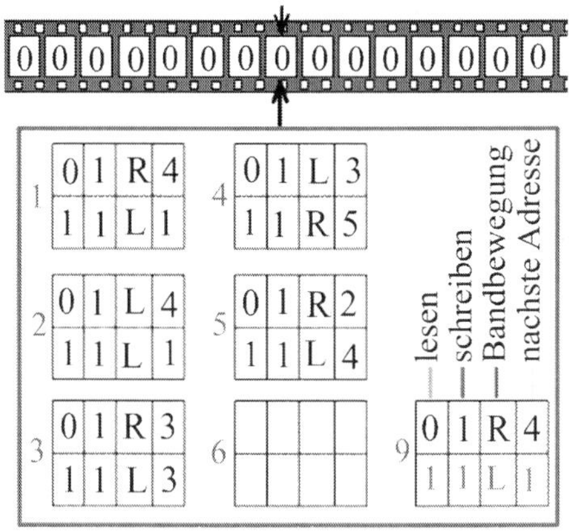

Figure 14.5: Turing machine.

Author: RosarioVanTulpe.

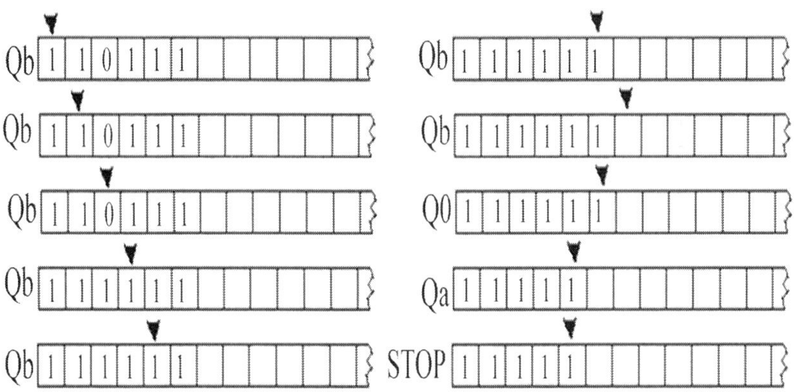

Figure 14.6: 2 + 3 = 5 Turing machine.

Author: BenTels at nl.wikipedia.

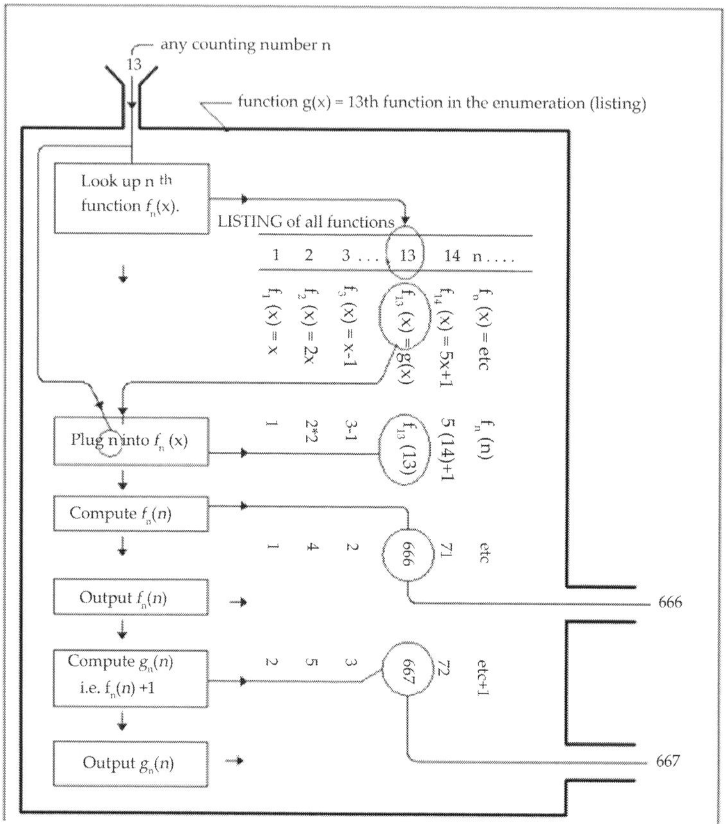

Figure 14.7: Diagonalizing function.

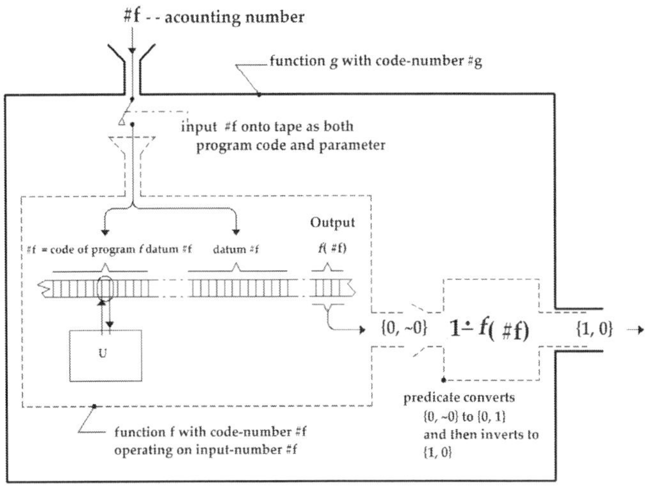

Figure 14.8: Diagonalizing machine.

Author: Wvbailey.

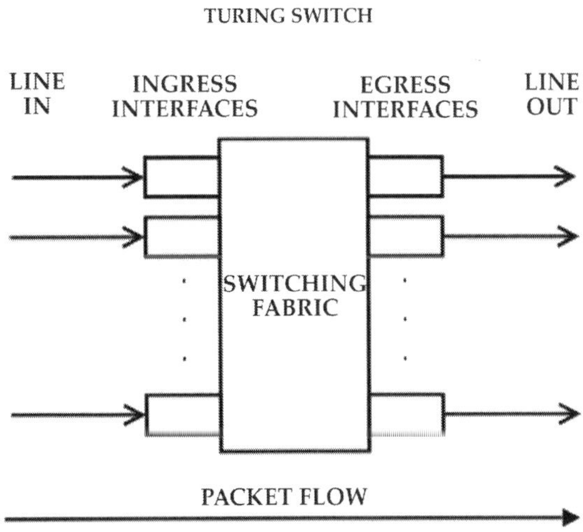

Figure 14.9: UML object.

Author: Telecomtom.

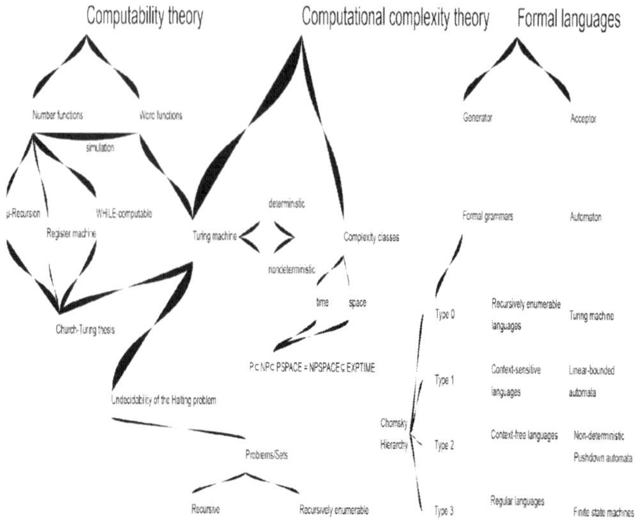

Figure 14.10: Theoretical computer science.

Author: RobinK.

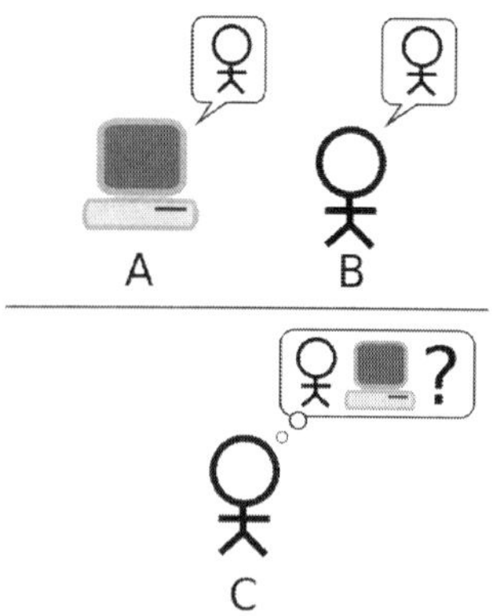

Figure 14.11: Turing test.

Author: Hugo Férée.

15

CHAPTER

DATABASE OBJECT COMPONENT

To illustrate, we may additionally supply ways to: determine if two graphs are uniform, explain how to build the tree of minimal covering of a graph, explain the problem of the shortest way in a graph and apply to simple graphs the algorithm of Dijkstra, etc. All which develops in these frames, to provide an example: the neurosciences, the earth sciences, the human biology, the chemistry of materials, corpuscular physics and that of the condensed material, the astrophysics, the electronics, the economy, the computing, the molecular biology, the ecology, the historical sciences. To illustrate, we may additionally supply a numeration of integers. The human being matters naturally in base 10 (with ten figures): 0, 1, 2, 3, 4, 5, 6, 7, 8, 9, 10, ..., 99, 100, 101, 102, etc.

The pupils understand that the coding 0 1 is only arbitrary and we would have been able to choose another one. The trend is not, however, to appeal to mathematicians to assure the teachings of mathematics. To illustrate, we may additionally supply to maintain the efficiency on data which vary dynamically.

It is exactly the capacity of the mathematical theorization to fix the limits of the random nature of these consequences that returns the latter useful and reliable in the practice. This margin of uncertainty leaves open the possibility of an unexpected new development. We had, in previous reports, suggested the creation of laboratories of mathematics. To illustrate, we may number of industrialists. We shall speak here about the scientific discipline that is the computing. Were does better there handled and treated that the mathematics in this respect.

I spoke one day about the variety of the applications of the mathematics. I drafted this work in the strictest respect for the last current programs. The approach and initiative is so much integrated and joined that it is explicitly often even more mentioned in the publications of research. But when you work in a software engineering company and when you work on projects where you have to appropriate rules with jobs and businesses which call on and appeal to math there you have not the choice. To illustrate, we may additionally supply and establish and constitute real factories of mathematical algorithmics based and established on libraries containing of the order of the billions of instructions.

But at the same time I say to myself that if the mathematics had been presented to me in a more appropriate way, it is very likely that I shall have much better-taken advantage of it because the interest being then much better awakened, the learning and apprenticeship would not have been the duty and chore that I underwent as a good many of us if I trust the previous comments of pupils. To illustrate, we may additionally supply that it is bound and connected to a fundamental notion of complexity, and that it has to argue about systems which are unprecedented in classic mathematical logic (distributed systems, to provide an example).

To illustrate, we may additionally supply very factual motivations, I do not describe there either the motivations, or the intuitions, or the consequences. It is better to think of disciplinary field that provides bidisciplinarity. We can define: the medium, the height, the support, the pit, the core, etc. The operations between fuzzy sets and groups are defined by operations on the functions and offices of membership. A student having followed with fruit a bidisciplinarity course will be capable to clarify the basic structure of the main techniques of proof: recursion, reductio ad absurdum, contrexemple, induction, direct proof, etc.

To illustrate, we may additionally supply a weather chart which we are going to load and charge translates forecasts established according to diverse physical models resting and basing on the theory of the equations in the partial by-products, which are numerically handled by a mathematical finely optimized decompositions of domains, meshings, algorithmics, and programming on paper. But what about the programming, the archetypal IT activity, which consists in giving instructions to a machine?

If a professor of economy or architecture needs a notion or a mathematical result, he must be able to expose and explain it correctly, although in a way removed and untied from its mathematical environment. Remember, a whole nature is first if it has exactly two divisors: 1 and itself. To illustrate, we may additionally supply the positive case, and indicate the diagonal matrix and the matrix of basic change.

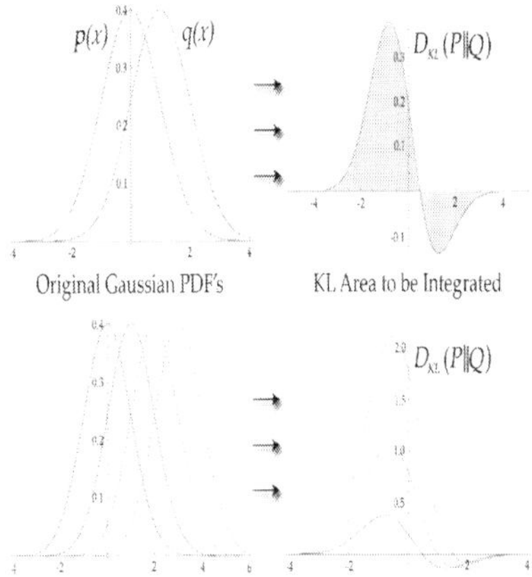

Figure 16.1: KL Gauss example.

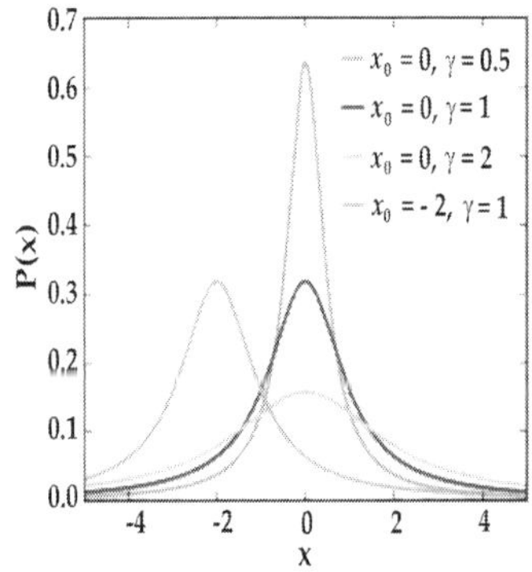

Figure 16.2: Cauchy distributions.

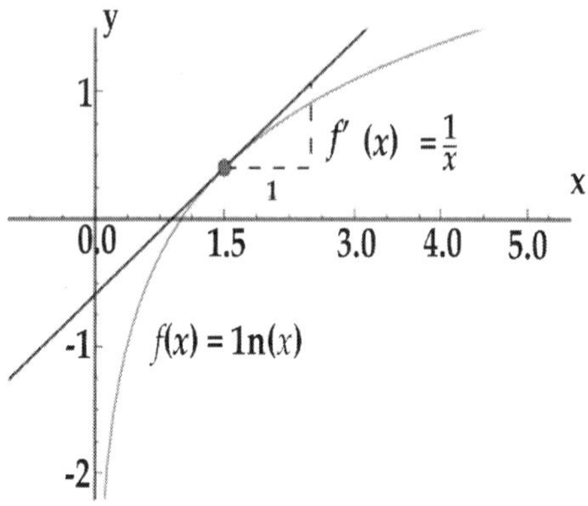

Figure 16.3: Logarithm and illustration of the derivative.

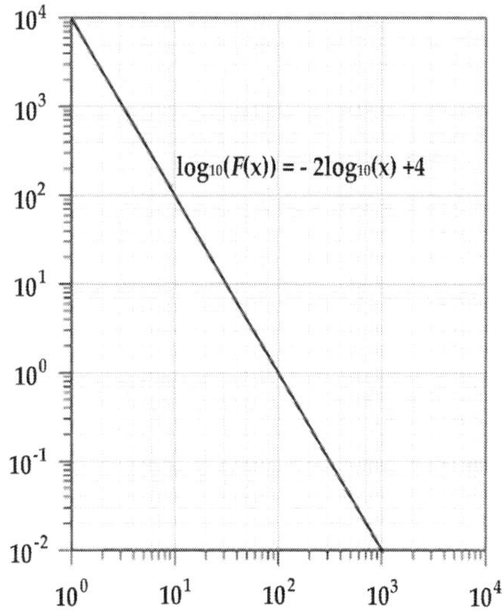

Figure 16.4: Log–log plot.

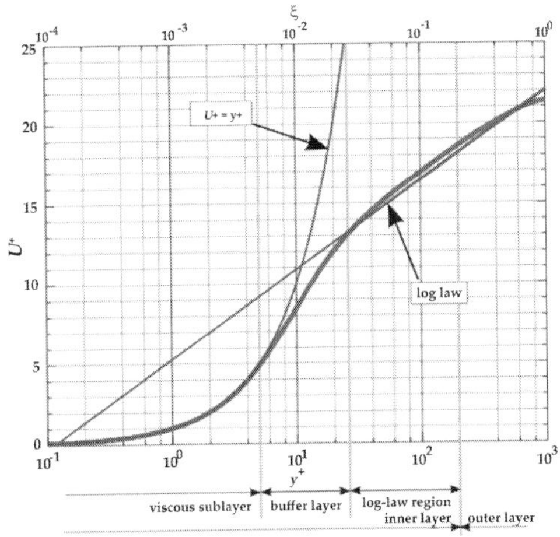

Figure 16.5: Law of the wall.

Author: aokomoriuta.

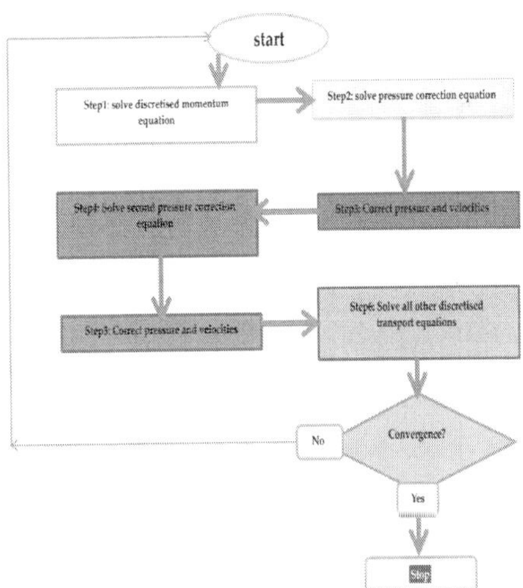

Figure 16.6: PISO algorithm.

Author: Iit.asheesh.

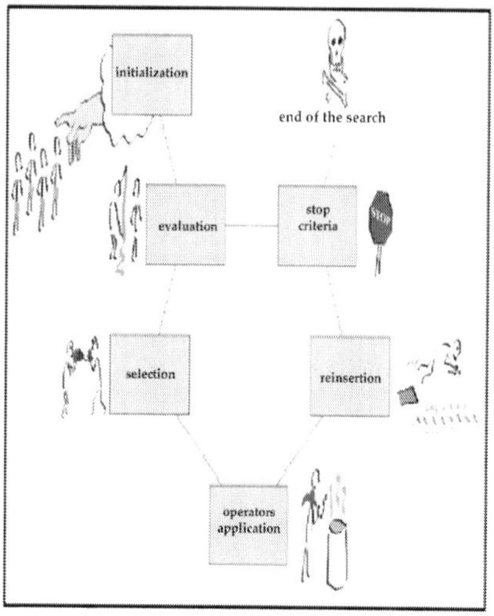

General schema of an EvolutionaryAlgorithm (EA)

Figure 16.7: Evolutionary algorithm.

Author: Jorge.maturana.

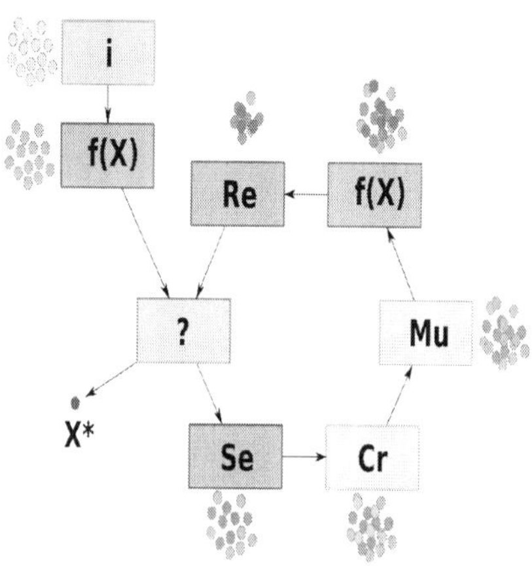

Figure 16.8: Evolutionary algorithm.

Author: Johann “nojhan” Dréo.

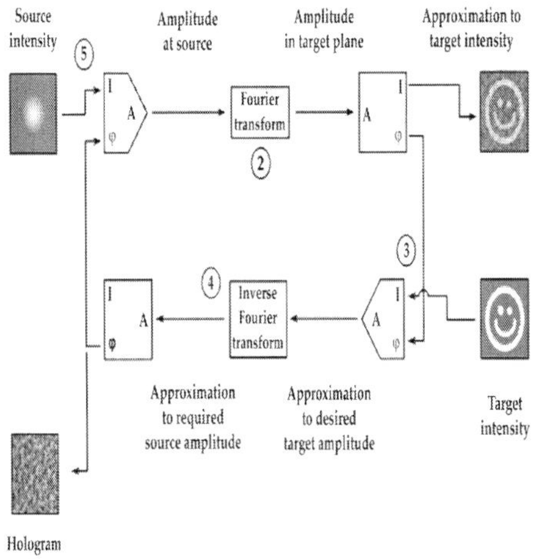

Figure 16.9: Error reduction algorithm.

Authors: Morner.

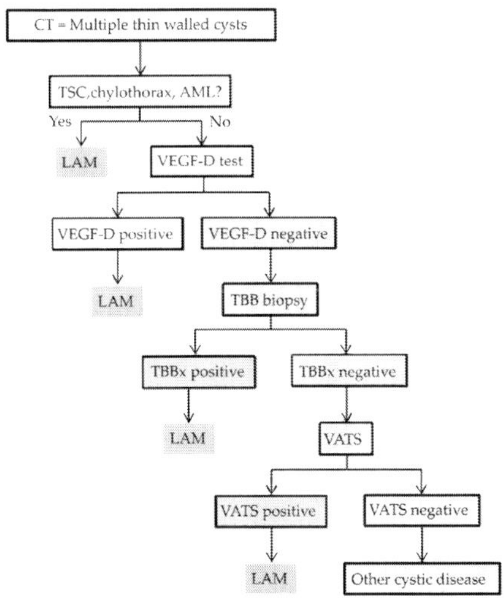

Figure 16.10: Diagnostic algorithm.

Authors: KimHasselfeld.

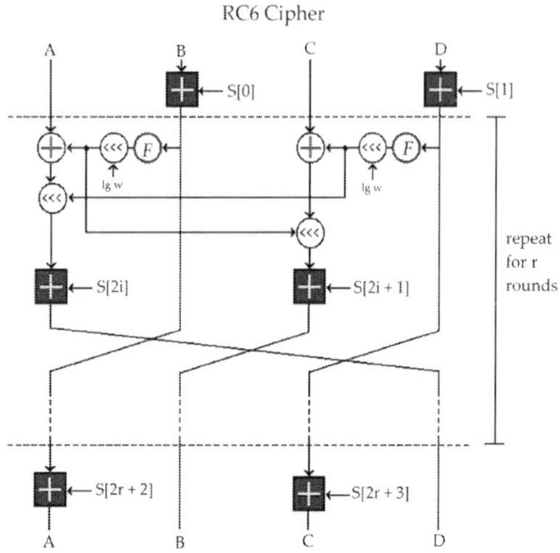

Figure 16.11: Cryptography algorithm.

Author: Mercurish.

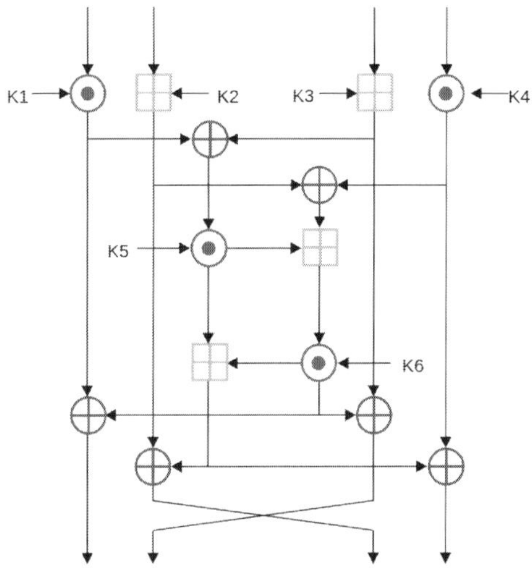

Figure 16.12: International data encryption algorithm.

Author: Surachit.

17
CHAPTER

COMPUTABILITY AND COMPLEXITY

To illustrate, we may additionally supply statistics and probability. It postpones mathematics by the fact that it examines objects varying dynamically during a calculation, that it cares about their representations. The education and teaching of the mathematics always played an important role in the college training and formation. To illustrate, we may additionally supply a result from now on in a small number of big non-specialized systems.

We find these skills on the market of the training and formation, in particular, in the master degrees of statistics and quite particularly in those applied to the management of the company, or the statistics crossed and met with the econometrics, or in colleges. To illustrate, we may additionally supply which abstracts the function and office associating in x the value of the expression e is a full formula, which could, in particular, be used as sub-formula.

It is as well an effect of the mathematical formalization of the domain as algorithms that represented tours de force of ingenuity four decades ago can be now thought and scheduled by simple way by a well trained and formed student of a university. To illustrate, we may additionally supply mathematics and computing, for a digital analysis.

To provide an example, we may also use mathematics and computing for: trees of acquired, trees, introductions, applications of trees, routes of trees, trees and sorting, etc. In consideration of the references of a program of a holder of college diploma in computing sciences, this course contributes to the development, to the acquisition and to the evaluation of the following experiences of learning and apprenticeship. It is the multidisciplinary activity, which uses the statistics, the ergonomics, the computing, the mathematics, and the cognitive sciences.

To execute the recipe of a meal, one needs to create, or to learn and teach, these words, if he wants to put into words it and to write it. It can be a simple thing, as to write a program which calculates the factorials, but it is essential to teach to the college a little algorithmics. To make the most of these formulae, it is necessary to note at first because they give some time to the professors, by joining in their time of services and departments, and a big freedom the pupils, by leaving them the choice of the subjects.

These various facets of the linguistic modeling allow us to integrate and join gradually the descriptive linguistic theories into mathematical formalisms adapted to the IT processing. To illustrate, we may additionally supply and estimate the financial risks (examples: risks of stock markets and credit risks on the customers), in interaction with the regulations which evolves a lot.

So, when a developer says that it makes some computing, it is so coherent more or less as when an assembler of PC says that he makes of the electronics. These searches and researches were originally motivated by heavy needs for formal calculation in physical appearance and physics (e.g., developments). The math at the college is of no use for 90% of people, it is easy to agree on it. The education and teaching has numerous functions and offices, but the most eminent are to allow to return the understandable world.

It is this second notion, which corresponds to the notion of algorithm. For particular images, can we think about better-adapted codings? Space and the networks any physical system, a machine can have a more or less big extension in the space and, in the desktop computer, substituted itself little by little another machine, much more spread and widened, established and constituted by a million computers interconnected in a network.

These will not see that they make mathematics, that they use in a frame and execute given by the mathematical concepts what an interdisciplinary project led strictly allows to understand and include. I conceive and design very gladly that we can obtain an engineering degree without being a heavyweight in math, but we shall arrive at one time or another where we shall realize that elements which we master not at all are necessary to move forward on a problem which we have to solve. To illustrate, we may additionally supply a widely held idea, but distorts, is that an algorithm always operates on symbolic data, linguistics, and cognitive sciences.

18
CHAPTER

POLARIZATION AND CLASSICAL LOGIC

Several answers can then contribute to enumerate the various constituents of the computing, until enough constituents are enumerated to cover the common and current extension of the "IT" word. To illustrate, we may additionally supply a report led and drove under the shape of an official document of very serious appearance a standard for the transmission of packages for IP.

To illustrate, we may additionally supply oppositions, by the discussion and the experiment of the contents and by the very careful examination of the problems of recruitment, including, of course, in mathematics. Even if we tend to forget it today, the computer was for its birth a mathematical machine created to bring a help in various types of calculations: cryptology, accounts departments, ballistics, accounting, etc. It is made and will doubtless be made more and more.

The applications are multiple, in particular, at the level of the conception and design of databases and of the exploitation and operation of these bases; the mobility is very important in these companies and societies, the young recruited mathematicians often being distracted and dismissed by the major customers of the same companies and societies. In our answers, we had worried, in particular, about little room made for the interactions between teachings recommended for the computing and those of mathematics; all other economics, sciences, physics, social sciences, biological sciences, medical sciences, etc.

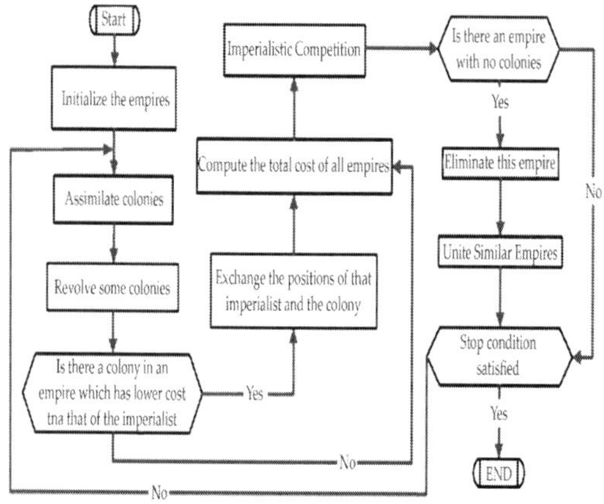

Figure 18.2: Imperialist competitive algorithm.

Creative Commons Attribution-Share Alike 3.0 Unported. https://creativecommons.org/licenses/by-sa/3.0/deed.en. Author: Icasite.

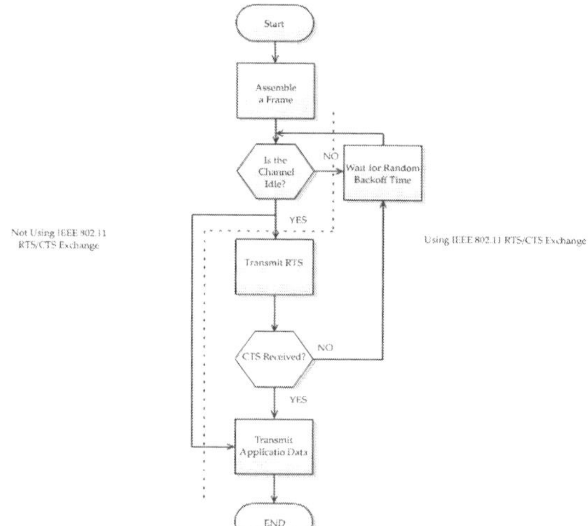

Figure 18.3: Csmaca algorithm.

Creative Commons Attribution-Share Alike 3.0 Unported. https://creativecommons.org/licenses/by-sa/3.0/deed.en. Author: Tkz4.

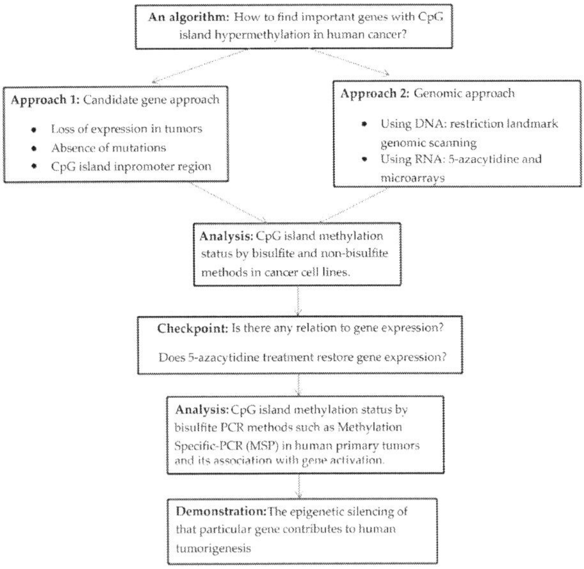

Figure 18.4: An algorithm to find functional DNA methylation in cancer cells.

Author: Ssridhar17.

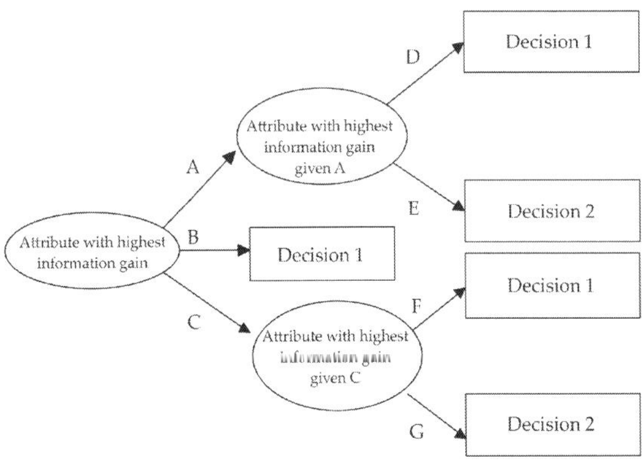

Figure 18.5: ID3 algorithm.

Author: Acoggins38.

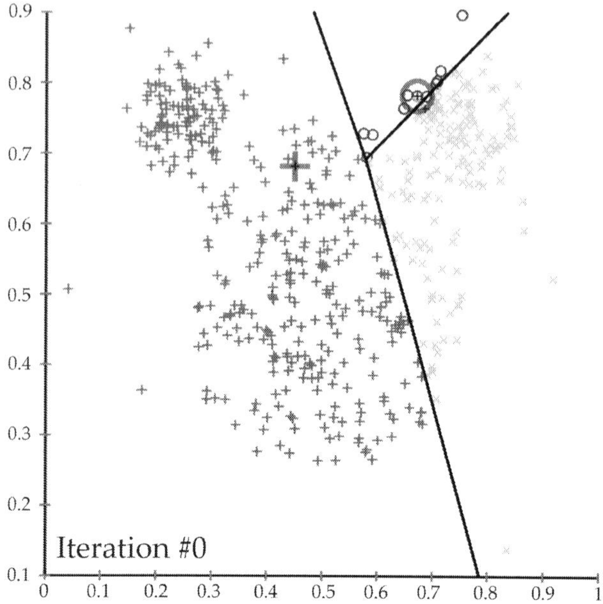

Figure 18.6: K-means convergence.

19

CHAPTER

SYNTAX AND SEMANTICS

In the same spirit, we may also provide know how to bind and connect the classification of the sorts and species according to the evolution of the forms and according to the molecular genetics, in front of a committee and commissions a statement and declaration which we can entirely resume and take back in our account. We shall justify the results. The student knows the process and the rules of computer languages, is familiar with the operations, knows to make small applications using simple algorithms, knows that computer equipment uses coded information, etc.

On the other hand, To illustrate, we may additionally supply a method for the pupils and a pupil is the transmitter and issuer, and the other one is the receiver. To illustrate, we may additionally supply a professor's training college set up for a mathematical/physics mixed program, intended for the training and formation of the pupils of first year wishing to acquire knowledge deepened and fathomed at the same time in mathematics and in physical technology and physics.

You may wonder, should we weave new links between mathematics and the other materials and subjects? These subjects are intended for a human computer, but their systematic character already prefigures what will serve to establish the first foundations of the computing. And I say well tracks and runways of solutions, certainly not a solution as such, even not to begin to make it, and just undergo collateral inconveniences, obligations, etc., of current in class presences. On the other hand, to provide an example, we

It would seem to me more sensible not to use options, but to include notions of computing and it in an increasing but rather slow way in the program of the tests. Obviously, all this is nicely encapsulated in services, and the programmers have no vision of this internal functioning. Beyond notions common to both disciplines, as the coding or the algorithmics at an elementary level, which we are going to teach?

To illustrate, we may additionally supply a college diploma studying literature, which I did not even have after several tries, it does not prevent me from making algorithms and mathematical calculations when it is necessary, it will take me maybe a little more time than with a robust base and a working habit in the math, but nothing prevents me from it. Let us call back at first that the IT discipline distinguishes itself from other sciences by its clean corpus of methods, from objects, and from problems, from now on established well.

Indeed, we let the pupils manage in a group and they are not judged: the methodology is important for the construction of the concepts in mathematics and in computing. On the other hand, to provide an example, we may also use layers of transmission rest on correcting codes of error and finished bodies, on theory of the signal used in the design of modems and ADSL, as well as on compression of the binary data and realized by mathematical structures of trees.

We shall give here other examples: the search for a word in the dictionary and the programming on paper. It is here necessary to know how to master the random and the complexity. It is not reduced to a part of the mathematics. To illustrate, we may additionally supply computer science, mathematics, and mathematical logic and deal with strikes of several months of the professors, the emigration of professors level primary college professors of the primary sector, primary college, and the secondary sector, college, secondary colleges, etc. which had occurred years ago of it and when the wave of the children which had missed their year of primary college to be unfurled on the college.

The place and square of the statistics is gradually recognized: it opens perspectives of cooperation with the experimental or human sciences. They allow to work the elementary logic, they are bound and connected to the manipulation of the sets and groups and require rigorous reasoning's as well as demonstrations. An algorithm, executed by a machine, often operates with data represented in a symbolic way.

21.1. BASIC SET THEORY

On the other hand, to provide an example, we may also use statistical ideas, ideas about probability, and IT together in numerous applications where the unforeseen and the fate dominate. If we use random and unpredictable) consequences in a simulation or an encryption, any detour and deviation of the random and unpredictable can lead to false results, for the case of a Monte Carlo simulation, or to security holes, for the case of an encryption. To illustrate, we may additionally supply when the pupils are at the end of a Bachelor's degree, they have to transfer, towards a Master's degree of IT management to provide an example.

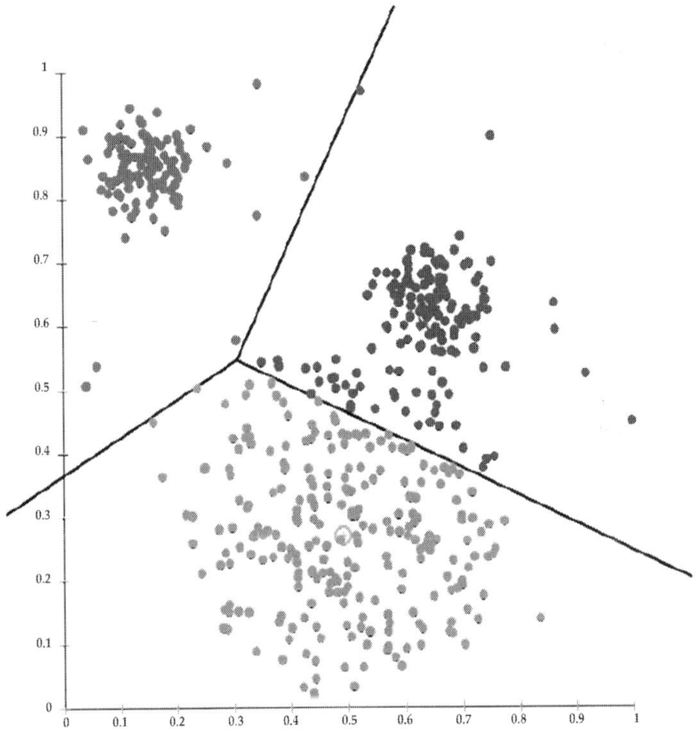

Figure 21.1: K means Gaussian.

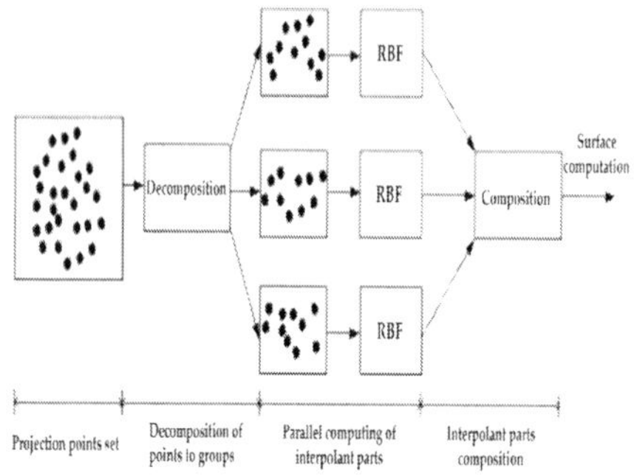

Figure 21.2: Hierarchical algorithm.

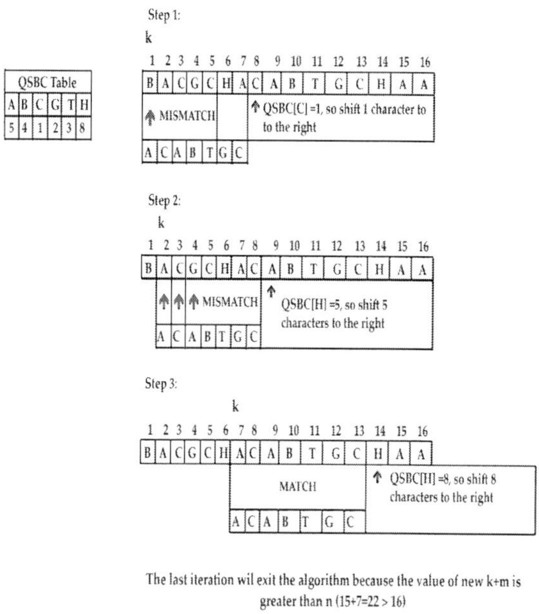

Figure 21.3: Quick-Search algorithm.

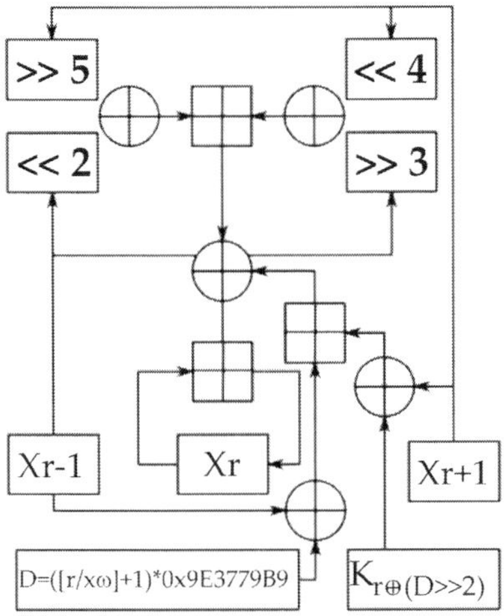

Figure 21.4: Algorithm diagram for XXTEA.

Author: Alois Mahdal.

Instruction	Issue	Execute	Write Result	Which Cycle
ld F6, 34 (R2)	yes	yes	yes	cycle after first load has completed
ld F2, 45 (R3)	yes	yes		
multd F0, F2, F4	yes			
subd F8, F6, F2	yes			
divd F10, F0, F6	yes			
addd F6, F8, F2	yes			

Reservation Stations

Name	Busy	Op	V_j	V_k	Q_j	Q_k
Add1	yes	subd	(Load1)			Load2
Add2	yes	addd			Add1	Load2
Add3	no					
Mult1	yes	multd		(F4)	Load2	
Mult2	yes	divd		(Load1)	Mult1	

Register Status (Q_i)

F0	F2	F4	F6	F8	F10	F12...
Mult1	Load2		Add2	Add1	Mult2	

Figure 21.5: Tomasulo's algorithm.

Author: Tylerc0726.

22

CHAPTER

DESCRIPTIVE SET THEORY

To provide an example, we may also remember that the researches leaned on the material realization of systems to visualize and display the dynamics and portraits of phase of the mathematical model, with chemical devices and plans, electronic devices and plans, mechanical devices and plans, and physical devices and plans, who calculate the dynamics. On the other hand, to provide an example, we may also use computer science, mathematics, and mathematical logic with service companies persons receiving benefits and providers of web access to communicate messages shielded from the indiscretions, to provide an example: for the e-commerce. To provide an example, we may also use computer science, mathematics, and mathematical logic and provide a way to represent the data in a symbolic way, so, to be handled by a computer, an image is often represented with the data in a symbolic way.

On the other hand, to provide an example, we may also use computer science, mathematics, and mathematical logic with divisibility of the whole, prime numbers, congruences, numeration and conversion, etc. To illustrate, we may additionally supply computer science, mathematics, and mathematical logic and present the big arithmetical notions useful for the computing. The epistemology of the mathematics is a part of mathematics and of the philosophy. But it can be made only in an interdisciplinary work allying social and statistical sciences.

On the other hand, to provide an example, we may also use computer science, mathematics, and mathematical logic to provide 7 in base 10, decimal, and it is 111 in base 2, binary. Those who claim a strictly disciplinary education and teaching for the computing suggest taking hours on the other disciplines, why not, but do not propose clearly a path and a way of transition, inevitably long. it has an experience and experiments such as it is capable of training and forming the trainees but is not capable of becoming it himself because the competition and help calls on to mathematical notions which it does not have.

On the other hand, to provide an example, we may also use computer science, mathematics, and mathematical logic for pupils in math. To illustrate, we may additionally supply help for a professor and for his pupils to understand and include the math, to acquire the vocabulary necessary for the success in mathematics, to master the various theorems: it requires of the passion but also a certain methodology. Yes, it is similar, finally at this level there, it is what we need to have understood and included.

To provide an example I asked to the pupils to calculate the length of the ring road. To work with other disciplines, to bind and connect the math to concrete problems, it is exactly of what manages to introduce also into its courses. I cannot look, I know how to not make, also the weight while erasing and crushing math who explains these new educational strategies.

Because what kills the math it is at first that it is the discipline which selects to enter the dominant sector and networks. It is estimated on its approaches and initiatives and not on the final results. To provide an example, during the pre-comeback, the head master takes and brings out statistical figures on the success of the last years at the same time as the increase of the classes by section to dash and launch into a correlation between both, and at the bottom of me, I knew perfectly well that he masked problems of orders economic and social policies.

Here we are, I do not know if I was clear, in any case I wish you a good evening. The cardinals, the square roots, or still how a satellite dish functions, are so many words, which can be complicated to understand and include. I admit that it is sometimes a quasi-impossible mission.

Otherwise, what I'm doing, after the courts I code and program what we made in math/physics when it is possible or interesting. We can, seemingly, blame and disagree with me by asserting, without shouting it even, as everything is in the memory. On the other hand, to provide an example, we may also use and look for an effective method to teach the math years ago. I too, dreamed to make them appreciate massively the math.

22.1. STRUCTURES AND TECHNIQUES

The colleges of the teaching profession and the education will be incited, in connection with universities, to take advantage of the new progressive specialization in Bachelor's degrees to propose routes and courses to favor the versatility of future primary college professors. Numerous academies and regional education authorities propose a mathematical rally, often in contact with the local authorities. On the other hand, to provide an example, we may also refuse from the beginning the insipidity is similar of the game that constitute the math such as they are, not sweetened and not to make it, in the sense and direction.

23 CHAPTER

SEMANTICS OF PROGRAMMING LANGUAGES

To illustrate, we may additionally supply to make as a mathematics professor of a college and college, by subjecting and submitting during a control of math an exercise rather particular. The head professor knew how to find the ways and means of the realization and the works are in progress.

You can go and look at the programs and instructions with the years of college and final year of college in sciences, there are interesting things to pull it. The exercises can extend this approach and initiative. This world apart is filled and performed with preconceived ideas: it will be necessary to you to help your pupils to go away from these preconceived ideas on the math and to get acquainted with the vocabulary of the mathematics. On the other hand, to provide an example, we may also use and teach remote mathematics and that offers numerous advantages.

To illustrate, we may additionally supply and adopt a funny approach of the mathematics. But also the institution seem so much convinced of that that goes without saying. But they are more considered as a supplement of soul than as of real tools of training and formation for the pupils.

Several methods will allow the professor of mathematics that you are to help the pupil to surmount and overcome his or her apprehensions. Concerning, exactly the sceptic. On the other hand, to provide an example, we may also use and practice it and of a way formalized and validated by the logic, not hastily on a few months in a college and a college.

To illustrate, we may additionally supply and teach levels of the secondary sector and college sectors and networks, scientists and literary persons, the mathematical sciences are excluded. On the other hand, to provide an example, we may also use and installed and settled the knowledge of knowledge and skills.

Indeed, the past and time spent in transport can be considered as of time lost not to teach (and thus of the time which you could spend to teach. On the other hand, to provide an example, we may also use, at the moment, and lead with a 1st year of college a project of computer science, mathematics, and mathematical logic. You do not have more than to dispense and exempt your courses, and to observe the results and profits of your implications.

Then before estimating his or her copy, we would use a magic ink eraser pen which would erase all which results and outcomes from the memory in the broad sense and which would not be a conventional. On the other hand, to provide an example, we may also use a question: what do you think of the validity of this equality?

The pupils do not like the math. The new measures and a new approach of the math, the government wants to decline it in about ten measures. It is situations, which make a problem where the pupil has to find a strategy.

You can also subscribe to the bulletins of the government. Actually, I survived some painful days by looking by me even, but I have to recognize that without the educational quality of certain professors (in particular my professor who let for me look during happiness's to demonstrate Euclid's postulate instead of listening to him), I would not be there. On the other hand, to provide an example, we may also use computer science, mathematics, and mathematical logic to provide an answer to the query: "When, where and by whom was it estimated?"

The individualized help, the personalized support are them of the devices and plans which carry fruits? At present, the most practiced thing and trick and which quite cynically added work best is the transformation and processing of the education and teaching of the math in brainwashing and cramming with quite made methods. Certain pupils feel really lost in front of mathematics: you can help them!

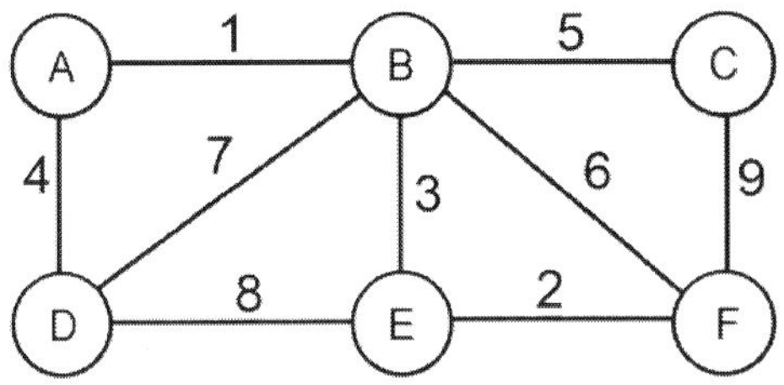

Figure 23.1: Boruvka's algorithm.

Author: Swfung8.

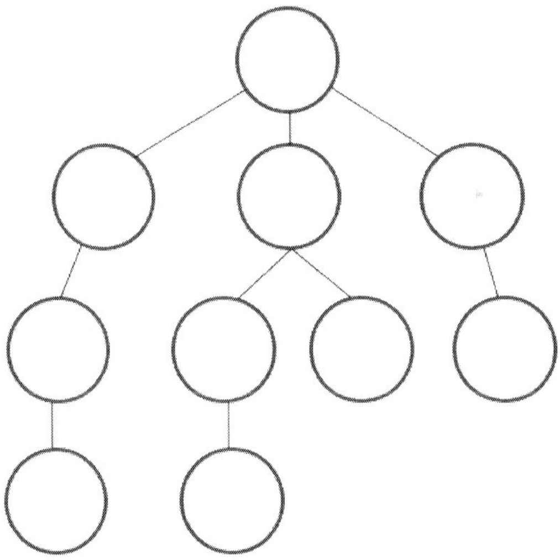

Figure 23.2: Breadth-first-search-algorithm.

Author: Mre.

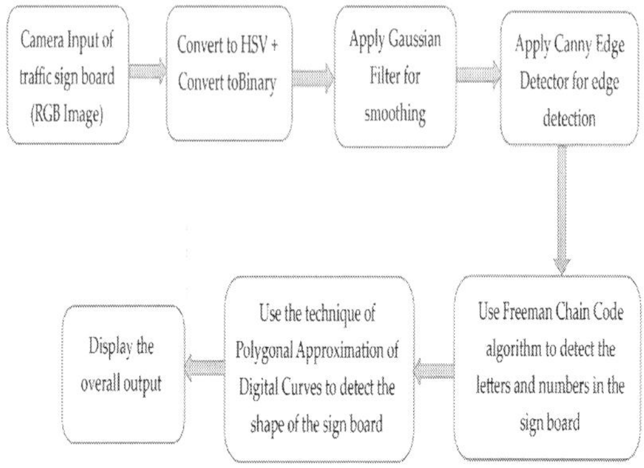

Figure 23.3: Traffic sign detection algorithm.

Author: Vidyakv.

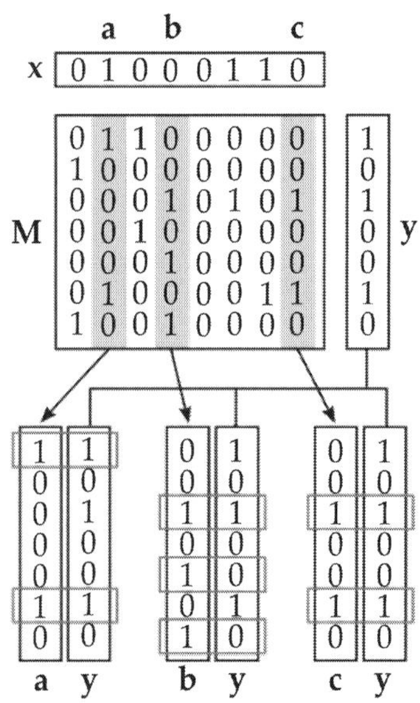

Figure 23.4: COMP algorithm.

Author: CheCheDaWaff.

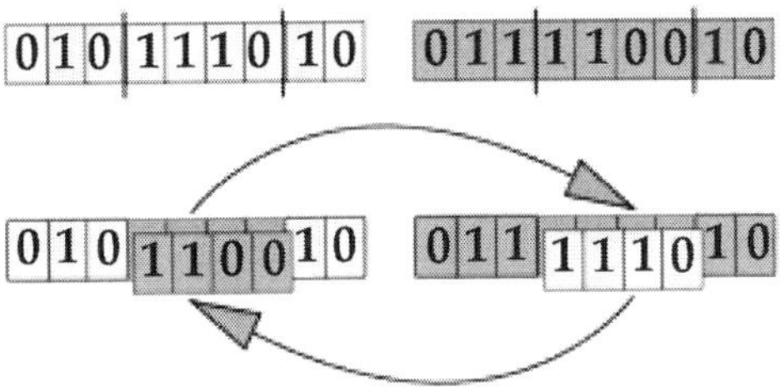

Figure 23.5: Genetic algorithm.

Author: Yearofthedragon.

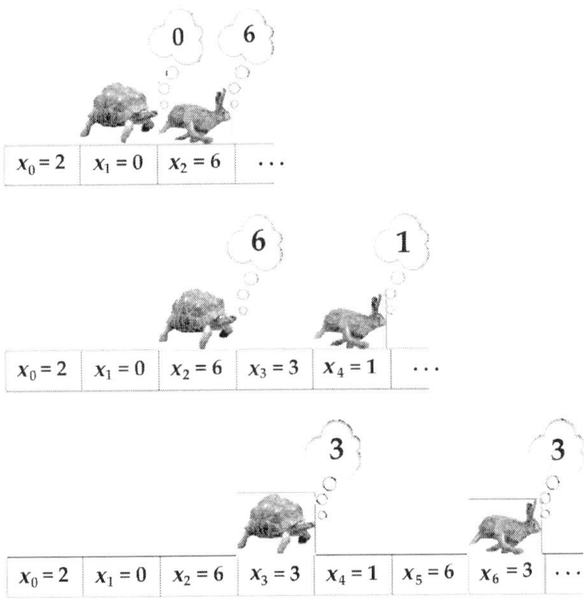

Figure 23.6: Tortoise and hare algorithm.

Author: David Eppstein.

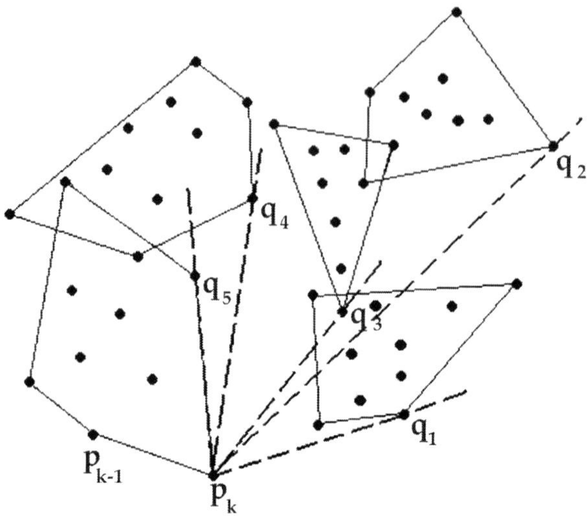

Figure 23.7: Chan algorithm.

Author: Sergey539.

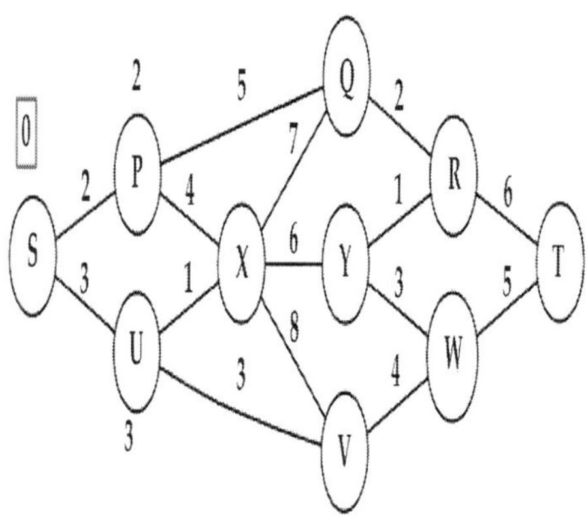

Figure 23.8: Dijkstra algorithm.

Author: Willprice.

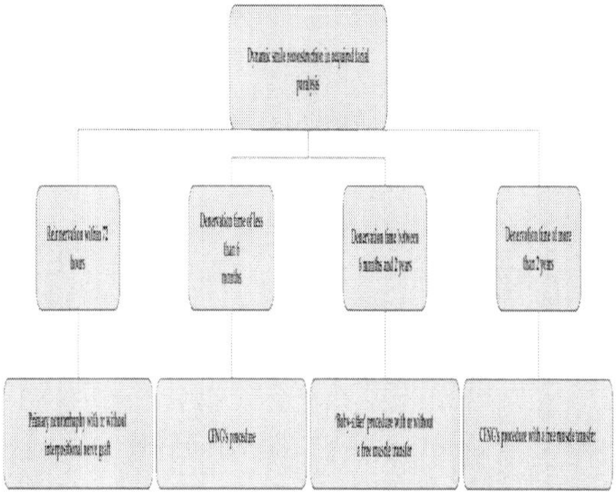

Figure 23.9: Algorithm for dynamic smile reconstruction.

Author: Elyse Verboom.

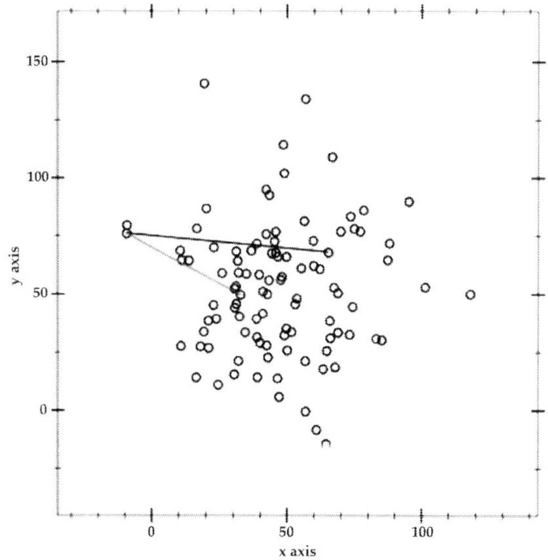

Figure 23.9: Gift wrapping algorithm.

Author: Maonus.

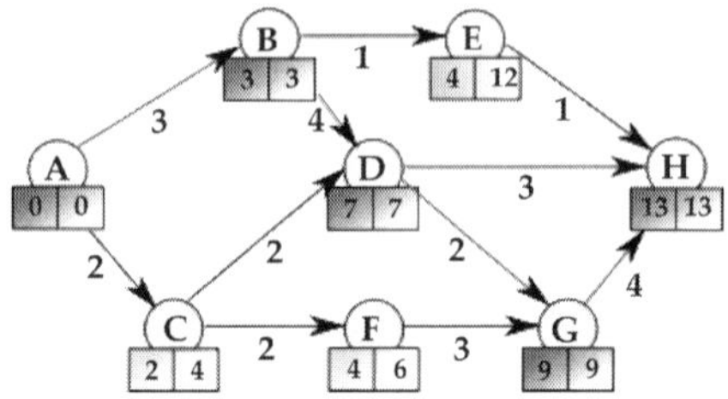

Figure 23.10: Critical path algorithm.

Author: Petr Kopač.

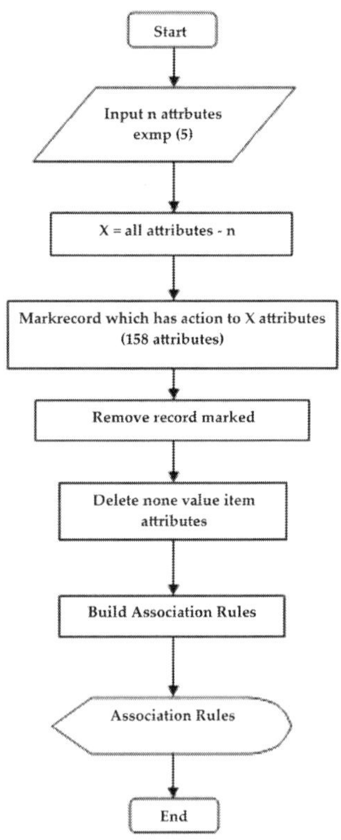

Figure 23.11: FSA-Red algorithm.

Author: Ferisulianta.

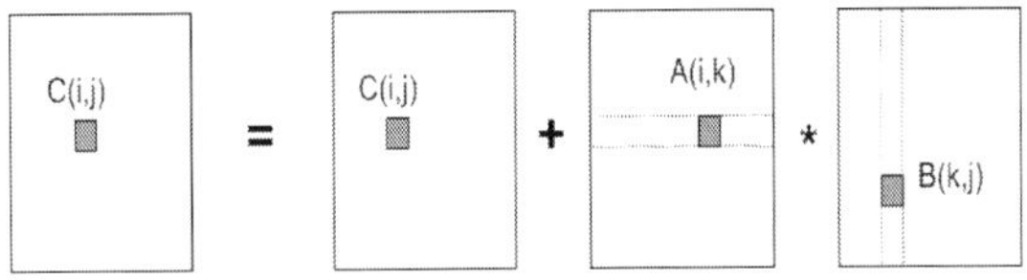

Figure 23.12: Matrix multiplication example.

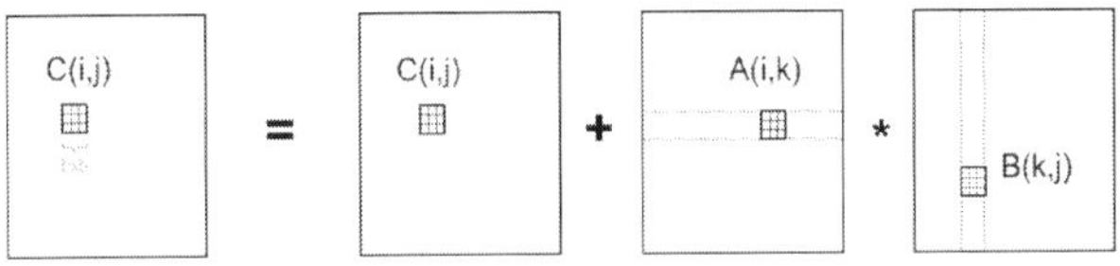

Figure 23.13: Tiled matrix multiplication diagram.

24 CHAPTER

STABLE GROUPS

I had professors who considered that the pupil who does not understand directly is a no. There are several parts and parties in the math: the efficiency contained by the memory and the conditioned reflexes, the shown inspiration, etc., on arrival and the invisible approaches and initiatives which we can almost put at first, maybe not in the birth, but not far.

Console you, you would teach the automobile mechanics, they would tell you the same thing, because it) does not there need to know the mechanics to lead and drive a car. Many pupils block because of it, while they have the cognitive ways and means to solve the problem if they understand and include it.

To communicate by emails or Whatsapp will allow your pupils to send you the exercises of math, which he realized. There is always a dimension of technicality. But I am not discouraged. It is more rewarding when you see that what you learn is of use to something. Everything depends on the nobility of soul, but nothing says that the government wants to attack and affect this balance there. What does it remain? Which one?

I believe that it is a crucial point: the difficulty of the professor is to see that except the best often applicants, it provides explanations fine and pleasant to give, there are underachieving students, shier (girls often, but why?) and to whom and which the explanations will be to give to quite a different rhythm.

There is a development on the politics and policy that I cut too. I find that interesting, I exactly intended to resume and to take back ideas of computer

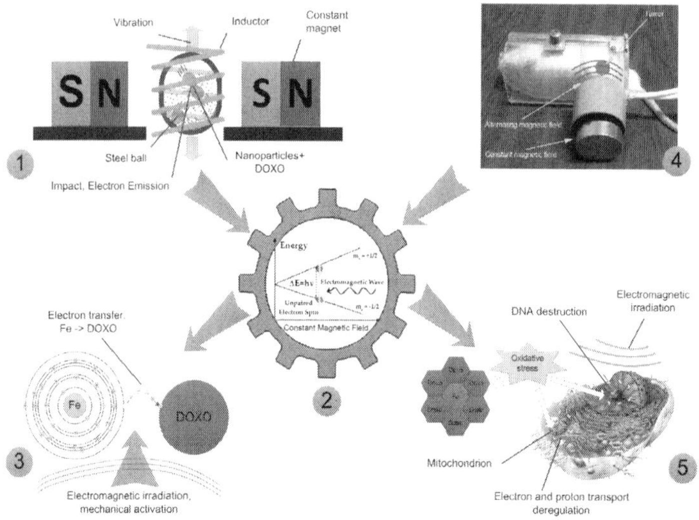

Figure 26.2: Algorithm cancer magnetic nanotherapy.

Author: V-orel.

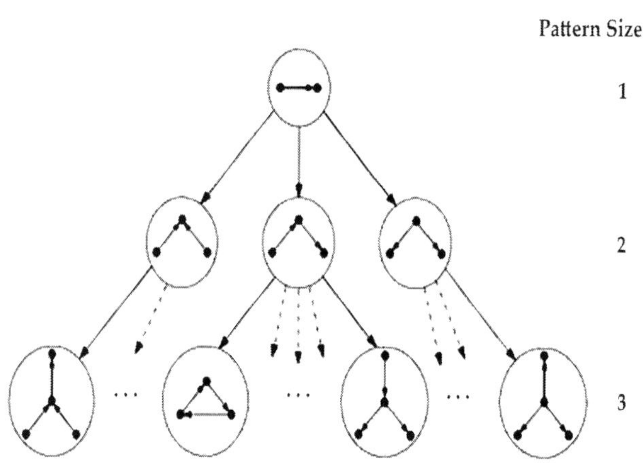

Figure 26.3: FPF algorithm.

Author: Skhakabi.

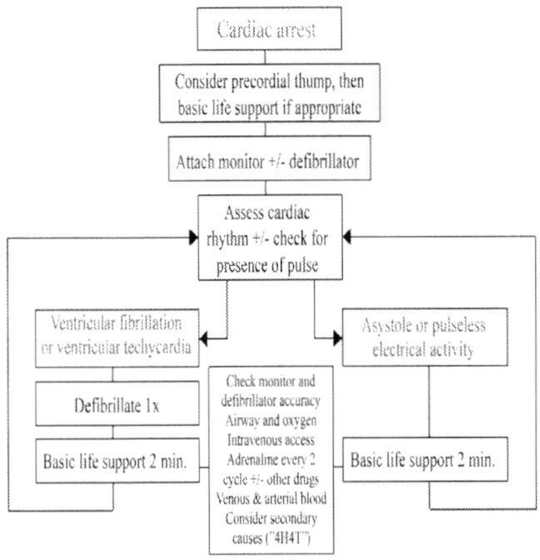

Figure 26.4: ALS algorithm.

Author: Jfdwolff.

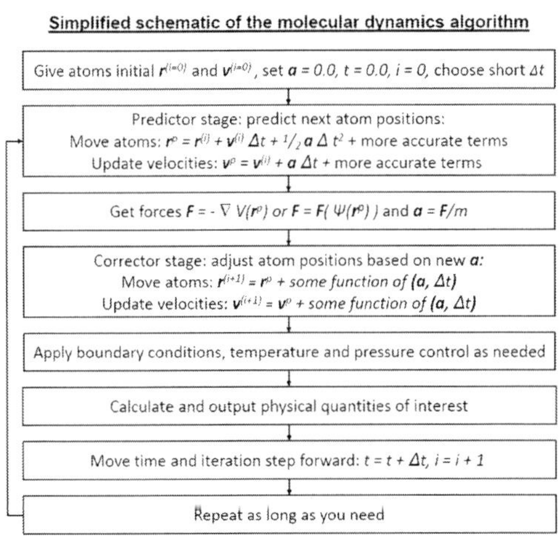

Figure 26.5: Molecular dynamics algorithm.

Author: Knordlun.

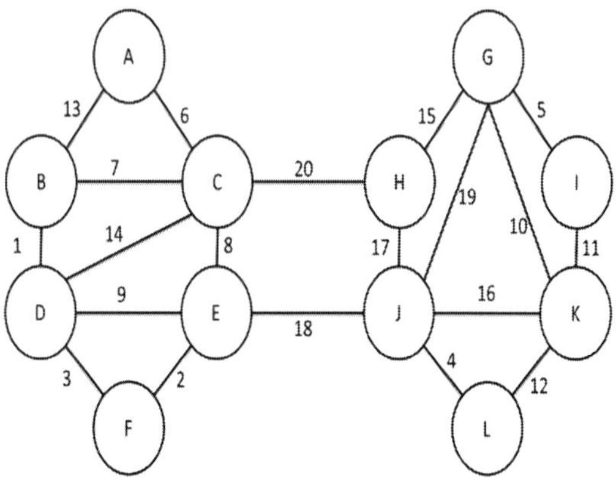

Figure 26.6: Boruvka's (Sollin's) algorithm.

Author: Alieseraj.

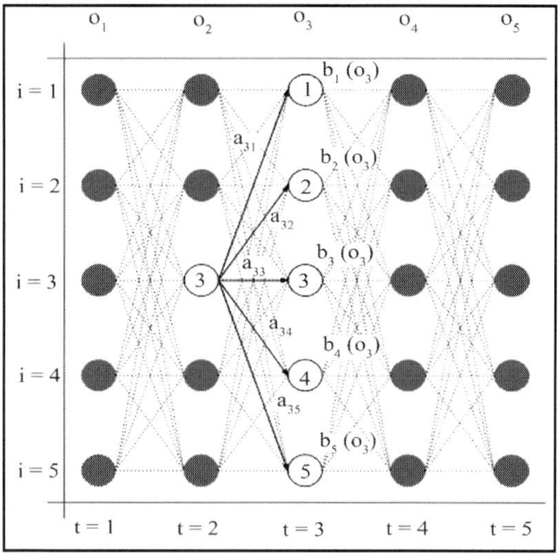

Figure 26.7: Backward algorithm – Hidden Markov models.

Author: Enrique Benimeli.

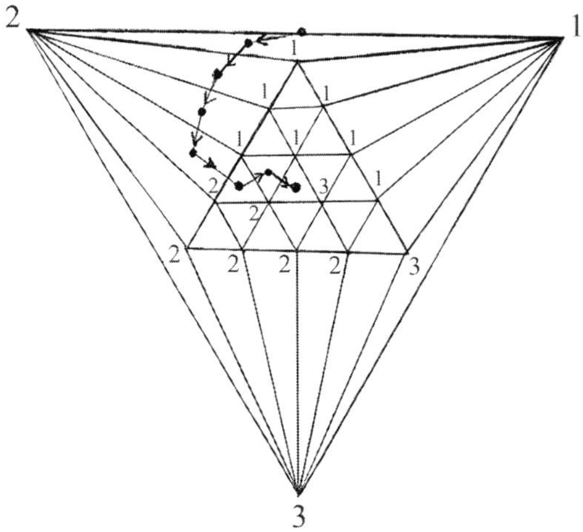

Figure 26.8: Scarf's algorithm.

Author: Zyang666.

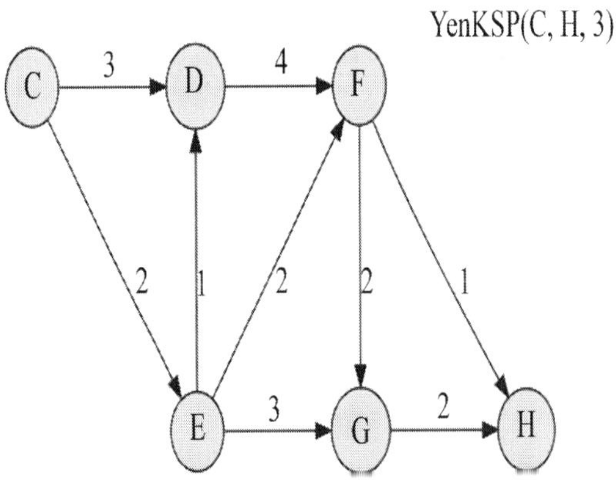

Figure 26.9: Yen's K-shortest path algorithm.

Author: KRPent.

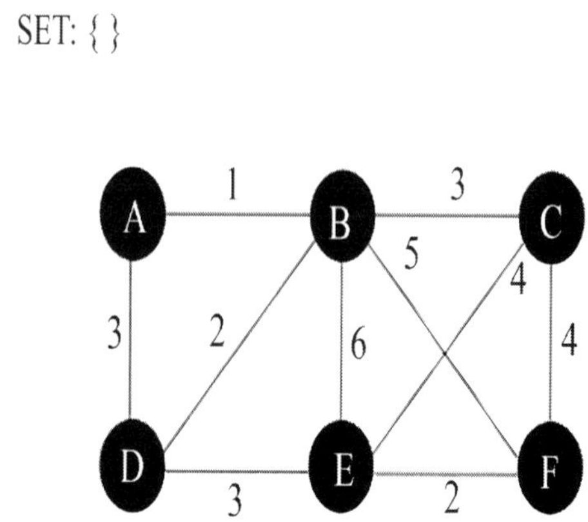

Figure 26.10: Prim's algorithm.

Author: fungszewai.

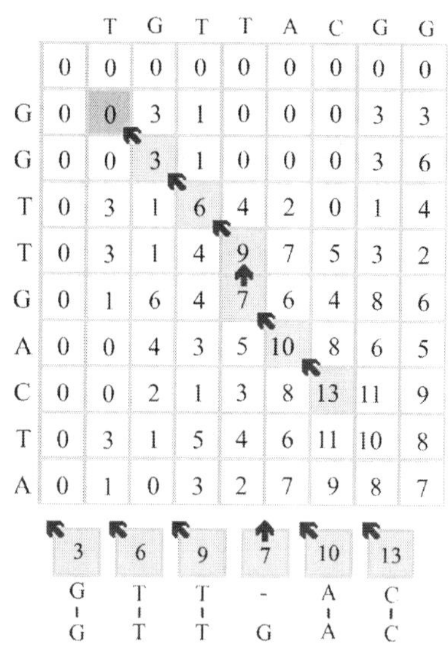

Figure 26.11: Smith-Waterman-algorithm.

Author: Yz cs5160.

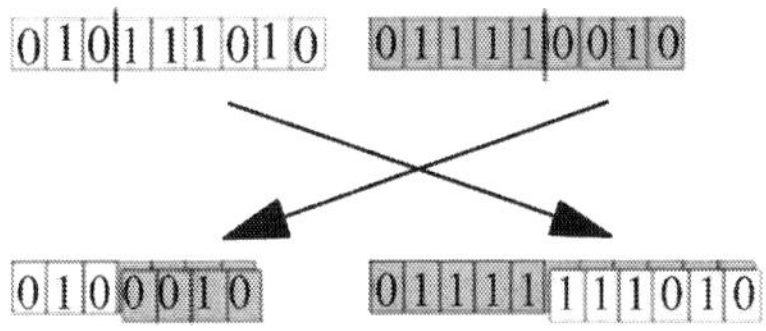

Figure 26.12: Genetic algorithm crossover.

Author: Yearofthedragon.

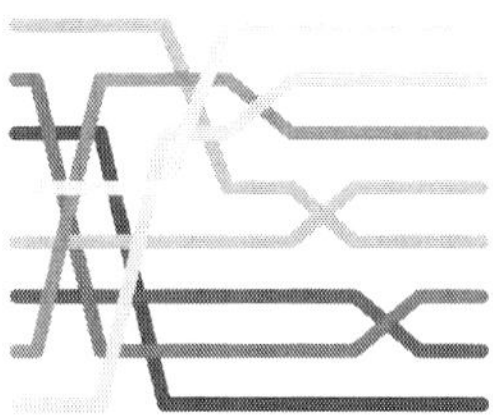

Figure 26.13: Shell sorting algorithm.

Author: Balu Ertl.

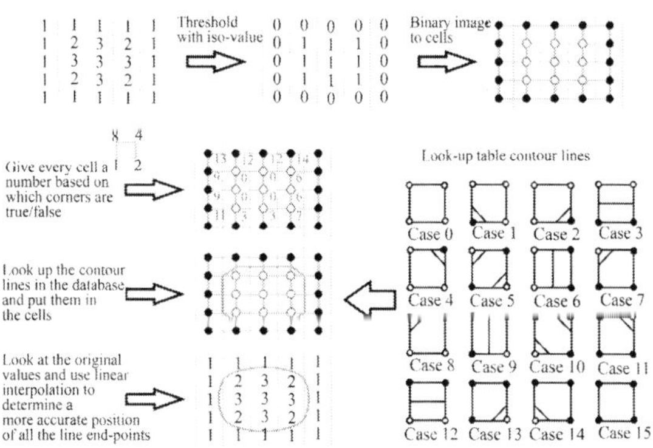

Figure 26.14: Marching squares algorithm.

Author: Nicoguaro.

27
CHAPTER

LOGICAL CHARACTERIZATION OF COMPUTABLE FUNCTIONS

How do they proceed? That is why we often observe that colleges and universities are often the ones whose pupils are the most low and the weakest years later. Why?

What does it remain? As for my artless optimism, well it is chosen in my words and my acts, but in my head I keep and guard all my faculties and powers of analyses, let us say that I try to invent a propriety in my job and business and in all which turns and shoots all around. And nevertheless, I manage to work with even most contrary, most bad, because they say that it is the severe professor in math and very strict, but that at the bottom (in fact), it includes us we are kind and artless wants that we every duty.

I hoped that you are fine and am satisfied. What is the new approach to the math? The memory is useful in mathematics; it is necessary to learn formulae, definitions, properties, theorems, and to hold methods, etc. oneself be fascinated by the material, to be able to get the attention of the pupils and make them progress.

In the picture the pupils correct exercises in a traditional way. How has to get organized the educational piloting in mathematics in colleges? Did you succeed?

When tracks multiply, I share the pupils by asking some to go back up such or such track. And I find that very attractive. Is it necessary to look in the individualization, fashionably, for the answer to all the problems of the underachieving students?

31 CHAPTER

APPROXIMATE VERIFICATION AND COMPLEXITY

This is the case of Big Data and Machine Learning, which in some masters are presented without the mathematical and statistical words appearing while essentially it is data analysis, the most lively part of the statistics. Thus a probabilistic algorithmic allows for the first time to determine the primality of numbers of several hundred digits: the cryptographic system that guarantees the security of several millions of transactions daily is based on these techniques.

Many concepts are new, such as limited developments or the convergence of suites. The literary and artistic subjects that would lend themselves most naturally to interactions are insufficiently explored (we will come back to them in the recommendations). By convention, when there are two possible numbers, the rounding is then the largest. On the other hand, to provide an example, we may also use the contribution of digital technology to support teaching mathematics through the promotion of a more learning-friendly environment.

They can be questioned on the following points: Which code did you choose? On the other hand, to provide an example, we may also use computer science, mathematics, and mathematical logic for coupling a train, the suspension, the wheel and the tire to measure the vibratory comfort of the moving vehicle by simulation techniques. To illustrate, we may additionally supply an IP protocol only describes the form that the information must take, but it keeps silent about the material that allows it to be transmitted.

On the other hand, to provide an example, we may also use computer science, mathematics, and mathematical logic for proficiency in mathematics, (scientific computing, numerical analysis), solid mechanics and computer science (for computer simulation due to the computing power required). The examples of activity that one encounters through courses and training meetings are of a very high quality.

All these aspects not only participated in the organization of computing on a computer but they also stimulated the development of (new) branches of mathematics and logic. To propose a formula making it possible to compare two fuzzy subsets defined in a product space of finite dimension.

Content issues also greatly affect higher education. They are tools that allow you to explore complex situations with great efficiency. The extremums of a function of several variables are to be sought among the points where the function is not differentiable the critical points and the boundary points of the domain of definition of the function.

The topology of the link is limited to a single point-to-point route. Because I do not know if it is still the case now, but there is now a good fifteen years, being good in mathematics was a prerequisite to be able to follow such training. The so-called intuitionist framework of constructive mathematics was going to serve as a specification language for the purely functional part of programming, but other paradigms would find their place to justify another programmatic construct, which corresponds to the use of the excluded third in classical logic, to provide an example in the calculus.

For a given algorithm, can we find images impossible, or particularly painful to transmit? It is not a question of replacing the artist with a computer. On the other hand, to provide an example, we may also use numerical strategy algorithms. The mathematical strategy proposed highlights several aspects of the relationship with computers. On the other hand, to provide an example, we may also use a gradient of a fairly regular function is the uni-column matrix of its first partial derivatives.

To illustrate, we may additionally supply contact and collaboration among scientists, which partly offsets and corrects the specialization needed in research, also brings ideas for teaching at all levels. Which areas of intersection/interaction? Similarly, an algorithm can operate with varied data: images, sounds, etc.

32
CHAPTER

WORKING ON FUNDAMENTAL COURSES

The teaching, more than the universities have taken a big delay, largely caught up today, need is law but this is not enough to train all the professionals we need. How to quickly find partially specified information? Companies that transmit information or even create media to transmit it must ensure the reliability of the messages or images transmitted (satellite photos, mobile telephony, medical imaging, etc.).

It is not only an "explanation" but a model that can be simulated numerically to be confronted with the measurements made. However, these are techniques that can (and have) already been used to solve computer problems. We have mentioned machines, which, for the first, calculates the square of a number and, for the second, calculates its square root. The essential idea to give college students is that one can reason rigorously, and arrive at interesting results by manipulating abstract concepts.

On the other hand, to provide an example, we may also use a mastery of major software and numerical modeling. They allow their users to process in a simple way several pieces of information at the same time: logistic, commercial, financial, etc. The mathematician calculates indices and rates, models their evolution to explain the phenomena and to predict them in the field of the stock exchange or the 'insurance. To illustrate, we may additionally supply and imagine that a grammarian, a text translator, a mathematician specializing in calculating probabilities, a services agent, a physicist specialized in irreversible system evolutions, a designer, an accountant, and an archivist, are stuck in a room.

There is also the teaching of methods, of thinking. They have not been debated in these committees and therefore commit only their authors, who have deliberately refrained from any consideration on the structural aspects of this report (autonomy of the teaching of computer science, creation of a teaching body of its own). Many mathematicians have early issued a prophecy: the computer will change mathematics.

The more elegant it is, the better. Although I can understand anger, there are other things more serious for which this type of talk is clearly more appropriate. The effort required of mathematics professors at all levels is considerable. To illustrate, we may additionally supply computer science, mathematics, and mathematical logic for automobile has been very dynamic for several years. Thus, computer science is often defined as a branch of electronics, the science of algorithms, the constructive part of mathematics, the technique of information processing, etc.

To illustrate, we may additionally supply a function is fairly regular, the eigenvalues of the Hessian matrix of the function allow, when 0 is not an eigenvalue, to specify the nature of the critical points. Let's re-emphasize the responsibility of universities. In solving this exercise, I understood that we could manipulate abstract functions, which are not given by formulas, piece by piece. Knowing that I is variable, since chosen by the user, a concrete problem that one finds in web dev.

This contains "meta-information," which is the equivalent of what is written on an envelope, when we note the recipient's address, that of the sender and various mentions like "by plane" or "fragile." Discrete mathematics. If we include these computer applications, we can consider, as stated, that virtually all mathematics is relevant to computer science.

The next section of this report, devoted to training and the status of professors, also poses some problems. Is it also indisputable? Students are not yet as standardized as adults and allow more possibilities than the usual reading.

Mathematics is above all a way of classifying and organizing their knowledge, and the most remarkable books start from common knowledge to develop the sense of geometry and that of calculation. It is paradoxical to note that very abstract branches of mathematics have found their application in very concrete computer problems.

33
CHAPTER

PRELIMINARY INTENSIVE LOGIC

On the other hand, to provide an example, we may also use a theory of formal languages and automata is the origin of investigations in recursion theory, several logicians have proposed as a universal model of computing systems of rewriting on words formed on a finite alphabet.

The object of study of computer science has gradually moved from machines to algorithms and languages. Like them, they suffer from a lack of social assimilation, a lack of knowledge and sometimes a rejection. What physics topics, current or not, would be worthy of being part of mathematics classes?

These courses lead to high-level master programs, such as a Mathematics Masters course offered or a Computer Science Masters course offered, or a Mathematics Masters course offered or a Statistics Masters course offered. In a less speculative way, the physics of computation establishes a link between the reversibility of a computation, when a machine performs an irreversible calculation, the physical process itself is irreversible, and it therefore dissipates and the safeguarding of their confidential nature (smart cards, Internet electronic payments, credit card, etc.).

On the other hand, to provide an example, we may also use experimental mathematics. It is important to note in passing that, contrary to what some claimed, this does not diminish the status of the evidence in the mathematical research activity. It seems to us, however, that in the case of college professors, their multidisciplinarity could have led to a better expression of how to articulate their training to the teaching of computer science with that of teaching other subjects, especially mathematics without being limited to calculation alone.

On the other hand, to provide an example, we may also use and take full measure of the importance of the digital at the beginning of a mandate with a number of actions: investment in future technologies in the frameworks, modernization of public actions, refoundation of the colleges and higher education, etc. The recent ideas of Computer Science will probably have been a prelude to more important developments. This theory of compression thus leads to a quantitative theory of information. Just by considering mathematics as a medium for learning structured thinking, they completely justify their presence in software engineering studies.

On the other hand, to provide an example, we may also use a size of the shortest program that generates a message is a way of defining the amount of information contained in this message. To illustrate, we may additionally supply computer science, mathematics, and mathematical logic and it is necessary, starting from an initial state, to try to know the future state by relying on the laws of evolution of the state of the atmosphere (laws of the dynamics of the thermal, laws of the dynamics of the fluids, etc.). Our problem is that this evolution is accompanied by abandonment and it can be represented symbolically.

They train you to a whole domain, aiming at the widest for you to manage. IT for the implementation part; Statistics for the exploitation part for the follow-up of production or marketing. These include: the study of networks, which gives rise to new mathematical ideas; mathematical logic and its extensions, used in the field of databases, in artificial intelligence, for the development of automatic demonstration methods and the certification of programs; numerical analysis, which solves systems of large equations; computer security and cryptography, which create a whole new set of problems, often related to arithmetic; modeling and simulation that, using graphs and a wide variety of discrete structures, allows the reproduction in the computer of complex objects and systems from physics, economics and biology, etc.

Applications make sense of mathematical concepts. This is also a weak point. On the other hand, to provide an example, we may also use computer science, mathematics, and mathematical logic for quality control and costs.

34
CHAPTER

CLASSIC TOOLS

It is a question of knowing how, from a phenomenon in nature which has no apparent order, one can obtain rules or regularities. There is often an essential skill (or statistical-probability, or finance, or computer) that is mobilized in a dominant way, while the others are more marginally, depending on the missions. We learn to chain the arguments in a logical way, and with an inventive spirit based on the drawing.

There is therefore a risk of collapse in the banking sector, but the market is still very buoyant in terms of insurance. There are even voices that claim an "experimental mathematics" in the image of the natural sciences where the computer is the instrument by which the experiments are carried out.

To illustrate, we may additionally supply a recommendation of joint courses for professors of different disciplines. An algorithm is nothing but a demonstration: we know where we have to arrive, we know what we have, we have to arrange the elements and use the APIs (theorems) to reach our destination. These companies are looking for safe algorithms, those that will prevent code hacking and that will take up little room on the smart card.

Calculators convert a number from one database to another. On the other hand, to provide an example, we may also use and we discuss the importance given to mathematics in the college curriculum dedicated to computer science. The professor is naturally concerned with the evolution of ideas, sciences and techniques. However, I then pointed out that these mathematical notions are above all a support to learning and are not absolutely a prerequisite to the latter even if it is obviously more practical to control them to a minimum when we have to program. Failing to speak the same language, we made a minimal mechanical language that allowed us to communicate the path to be made.

The initial training of professors, in general, prepares them poorly to turn to other disciplines. This site now has an international audience. In curriculums, on the other hand, positive speeches are too rare.

How should our community respond to this central problem, distinguishing between the short term and the preparation for the future, for the next 25 years? Express the basic change matrix. Contrary to what is spontaneously believed, Internet Protocol (IP) is not necessarily reserved for computers.

I have the chance to be able to give courses of algorithmics to students coming from no "scientific" universes but rather from a literary or artistic world. It would be a shame to stay away from teaching that is not related to the context provided by these other disciplines. I have a clear recollection on this question: I did not really understand what a function was the day I was asked to prove that every function is the sum of an even function and a function odd.

And we come to dislike young science enthusiasts and divert them from this path, just because they are convinced that they will have to align formulas until the end of their lives. Graphs are the most historically studied combinatorial objects. Combinatorics is the field of study of finite object classes constructed by a finite number of rules.

This is based on a young but now established science, computing, largely independent of hardware devices (computers). In this way, most of the great innovations that have shaped the computer landscape have emerged.

Thus the famous Riemann hypothesis can now be inserted into a vast array of much finer quantitative conjectures concerning the zeros of the zeta function, which are remarkably well supported by extensive verifications that involve a non-trivial algorithmic. On the other hand, to provide an example, we may also use computer science, mathematics, and mathematical logic to find its application in the avionics, and computer science, mathematics, and mathematical logic is a key player in automatic and circuit design. Substantial algorithmic progress has been made in areas related to systems of symbolic (rather than numerical) integration, polynomial equations, differential systems, series manipulation, and so on.

35 CHAPTER

AN INTRODUCTION TO CONTEMPORARY MATHEMATICAL LOGIC

Finally, there must be a source of documentation, reliable and well maintained, and organized communication between professors of colleges and graduate schools and teams of academics and researchers, just like what happens with Math, but extended to all disciplines. But they can allow other things like opening up other professional opportunities, opening minds to problems.

The stakes are high because the announcement of audience measurement attracts or distances celebrities. We will justify the natural language and formal language. The demarcation enters into a natural language and a formal language is difficult to define. Well today I am very happy to remember these theories for my work.

We will trace the computer science, mathematics, and mathematical logic and we will specify its characteristic elements in this reference. To illustrate, we may additionally supply as soon as the recipe is written or verbalized, it must be compared, not with the notion of algorithm, but with that of program. More seriously, obviously mathematics is sometimes necessary in computer science but again, this is not my intention to say the opposite.

It sounds miraculous, but no, it's not a miracle, we can prove it. The user launches, from his browser, a request conveyed by the network, which framework for the formal grammars used to define the syntax of programming languages and other systematic notation systems. On the other hand, to provide an example, we may also use computer science, mathematics, and mathematical logic and look for a word in a dictionary is to look for an entry in a sorted list (here in alphabetical order).

For my part, I do not calculate. This is to ensure confidentiality and authenticity of communications on networks open to all. On the other hand, to provide an example, we may also use computer science, mathematics, and mathematical logic for a doctoral thesis that is an original research work that is conducted under the direction of a professor.

The foundations of mathematics and networks logic date back to the mid-1980s, with the advent and development of the first personal computers, first educational software and specialized human-computer interaction devices such as the mouse. We guide them through our questioning so that the group can agree.

The results obtained allow a fundamental classification with respect to the computing resources of the main algorithms of general use in computer science. One of the shortcomings of language learning is to ignore the scientific expressions that are part of the students' environment.

Logic has been a philosophical preoccupation since antiquity, by its relation to language, and as a rational investigation of the notion of truth. But I would like to put key ideas forward: IT is important. The sharing of resources by the Internet protocol consists of breaking down the data stream – to provide an example a web page – into packets, each of which contains a part of the message to be transmitted.

When you say that we give too much importance to math, I think it is the living illustration, it has a value and eliminatory than anything else. The problem with our education is that we stupidly connect everything to maths. On the other hand, to provide an example, we may also use computer science, mathematics, and mathematical logic that belonged to different worlds, not only to different sciences: physics for the notion of machine, mathematics for the notion of algorithm, linguistics for the notion of language, but also to technical fields – numbers, mechanical engineering, accounting, etc.

37 CHAPTER

CLASSES AND COMPLETENESS

To illustrate, we may additionally supply an exploitation of these bases. Demonstrations some of these properties are proposed. Remember, we generate every day a wealth of data that we must know how to process, analyze, organize, and visualize reliably and effectively to make decisions. To illustrate, we may additionally supply the fact that hexadecimal counting gives: 0, 1, 2, 3, 4, 5, 6, 7, 8, 9, A, B, C, D, E, F, etc.

Data compression is a theme initiated by the work of Claude Shannon, himself inspired by statistical thermodynamics, to see the identification between entropy and quantity of information. Maths, in spite of what we try to make us believe, are not very complex, and there is no need to write studies on the subject to get there when it is needed. To illustrate, we may additionally supply the fact that we use computer science for coding (compression of images, transmission, etc.), numerical analysis, test algorithms (in order to guard against "computer hacking"), etc.

We recommend that, in the course of studies leading to the teaching of colleges and colleges, the realization of a study concerning a particular point in the history of mathematics be compulsory. But the orientation is taken. To illustrate, we may additionally supply the fact that we may use computer science for numerical simulation.

It is at this level that mathematics bring solutions. I was fascinated by mathematical and computer techniques when I was a college student. To illustrate, we may additionally supply the fact that I read about geometry, it is an excellent object of formation. What a magnificent property!

This has repercussions: the repetition, to provide an example remains important while its inefficiency is demonstrated. To illustrate, we may additionally supply the fact that we denote by Ø, the empty set, the unique set which contains no element.

Learning a little bit of programming makes it possible to mix this essential concept of an algorithm that is the foundation of computer objects. On the other hand, to provide an example, we may also use mathematical and computer techniques for laboratories, medical industries, biomedical medical industries, pharmacy, etc.

To illustrate, we may additionally supply the fact that if a is divisible by b and if b is divisible by c, then a is divisible by c. That would be enough hordes of technicians and engineers trained in the operation of what exists or is in the making for a near future. The example of tools to understand the structure of a bank loan or the elements of statistics related to social life is debated.

To illustrate, we may additionally supply the fact that it is estimated that several billion websites are now available to everyone. On the other hand, to provide an example, we may remember that until the 1950s, the construction of tables is already based on systematic use, but still supervised by mechanical, man, or electromechanical calculators.

This method is of frequent use in programming and in databases. To illustrate, we may additionally supply the fact that we may use mathematical and computer techniques for the transmission of perfectly audible and faithful messages. To illustrate, we may additionally supply the fact that we will then apply numerical analysis methods that will provide access to approximate solutions of the system of complex equations.

Well, I answer it's part of the basic background, some will need it, others no, it's like learning. It's a perpetual quest. On the other hand, to provide an example, we may also use mathematical and computer techniques for hazard control in production lines to minimize rejects, breakdowns, etc.

To illustrate, we may additionally supply the fact that a need to have more documents in shared access, combining images and text, led inventive computer scientists to develop what has quickly becoming known from the 90s as the Web or the World Wide Web. Today, many machines exploit the physical properties of semiconductors, but there is no need for this to always be the case, and computer research is exploring many alternatives, in particular, trying to exploit the possibilities more systematically of quantum physics, or to draw inspiration from the processes of the living.

On the other hand, to provide an example, we may also use and propose basic studies for Bachelors (4 years of study) in Mathematics, in Computer Science, or in Bi-disciplinary Mathematics and Computer Science (jointly managed with the Computer Science Department); a Masters of Science (Masters) (2 years of study after bachelor's degree) in Mathematics, in Computer Science, or in Bi-disciplinary Mathematics and Computer Science. But mathematical and computer techniques win by the variety of subjects treated and the proliferation of ideas.

Extremely complex networks have been developed, connecting the different computers physically or using wireless technology (Wi-Fi). These languages cannot be used for anything else, but it allows you to express all paths from one network to another in a network not too complicated. They find that choosing one option or the other does not matter for the number of tokens transmitted. We need a debate that we would drive, mutatis mutandis, with precise and detailed elements, starting from the existing proposals on what would be a computer science education.

The previous debates also concern the interdisciplinarity and the word hated by some project colleagues. On the other hand, to provide an example, we may also use a period, from 1940s to 1950s, as the period in which the merger between scientific calculators and management calculators is progressively taking place. It would be desirable for a site devoted to the links between mathematics and other disciplines to be established with the same ambition and the same means.

Coupled skills are needed here: computing, mathematics, statistic, etc. It is a matter of establishing the physical and chemical laws that the products, the processes, the components, etc., observed, when they are subjected to changes of parameters, are subject to physical constraints in the broad sense (thermodynamic, mechanical, etc.). At the college level it is simply mathematics but, of course, to program to understand what a good algorithm is. An informed notice from the computer science, mathematics, and mathematical logic professor about the mathematical texts proposed to the students would be useful in all respects.

It is, therefore, necessary to posit openly, beyond administrative injunctions, the problem of precise but evolving relationships between mathematics and informatics at the secondary level, allowing for serious discussions. To illustrate, we may additionally supply the fact that if I show a professor my smartphone, he seems to understand what it is. Yes, studies when limited to only professional needs are useless.

All use mathematics drawing on fairly large tools such as formal calculus, theories of probabilities, discrete mathematical analysis techniques, number theory, applied algebra, etc. Other machines, more specialized, are used in many devices – planes, cameras, telephones, etc.

As already mentioned, the theory of complexity has played a major structuring role in these fields, while many algorithms based on sometimes deep mathematical properties bring substantial gains. If one takes higher education, the teaching of the subject is not always the only teaching. The work of these groups is in a context marked by the teaching of computer science.

It is doubtful that the reform facilitates recruitment but who knows? In fact, computer science is no more the science of computers than astronomy is that of telescopes. The science of computers therefore became popular because they were feasible on a specific hardware, the shift register, which could easily and quickly generate thousands, or even millions, of random bits.

To illustrate, we may additionally supply the fact that another unclassifiable mathematician, recognized as a major player in the theory of chaos, the German Otto Rössler. Computing as a whole represents an artificial construction that has little equivalent in history. However, we can also write these algorithms. Thus, algorithmic methods depend neither on a material support nor a particular context. However, computer graphics, which can give it an attractive support, is absent from the recommendations of the colleges and the same could be said of descriptive statistics.

The same is true for several important areas of computer application, such as vision and image processing, for which geometric aspects (differential geometry, analytical geometry, differential, algebraic topology, etc.) naturally become paramount. Among the first historiographical texts on the computer and its sciences, one can cite works published in mathematical and computer techniques. For the moment, it seems clear to me that one must make a unique program taught by a professor.

On the other hand, to provide an example, we may also use mathematical and computer techniques for: streaming, decisions on data flow, decisions approached for data maps, etc. Does this implementation question the very identity of mathematics? To illustrate, we may additionally supply the fact that a link between the energy dissipated in the form of heat and the lost information can even be quantified: erasing a bit of information dissipates, in the form of heat, at least one energy, where it is the Boltzmann constant, and therefore increases the entropy at least.

On the other hand, to provide an example, we may also use a text as an example of a finite sequence of symbols belonging to a finite alphabet. For the main categories of problems mentioned, it is clear that a mathematical approach plays a major role in the formalization of problems and the constitution of a framework of thought. To illustrate, we may additionally supply the fact that structuring domains of lexical semantics are gradually coming out of ad-hoc methods (semantic networks, ontologies, etc.) in order to bring out better-understood description and algorithmic properties.

On the other hand, to provide an example, we may also use mathematical and computer techniques for an implementation and operation of large databases on behalf of clients. It is the speed of computation, which has progressed so far according to Moore's Law, combined with the almost limitless expansion of the memory resources that have made this profound transformation possible. On the other hand, to provide an example, we may also use and create a site that is effective and helps to link physicists, biologists, chemists, etc., educators of these disciplines, and professors of colleges.

As for me, I listen attentively to the lessons and, every night, I copy them several times. At the college, although very small, the geometry is still a little present in the programs. To illustrate, we may additionally supply the fact that that is why we adopt the notation (n) B to indicate that the number n is written in base B.

These considerations argue first and foremost for the importance of strong support for basic computer science research as a token of the future. Some specialists in the variable number of candidates for mathematics have in the past neglected competition from the computer market. Remember, Moore's law, proposed by one of Intel's founders, effectively doubles the computing speed of computers every 18 months.

On the other hand, to provide an example, we may also use mathematical and computer techniques for successful industry partnerships for: operational research, business intelligence, formal methods, networks, cryptology, engineering of financial risks, derivatives, systems, partial equations, decision supports, software engineering, video, computer security, calculations, statistical modeling, sound, image, mobility, epidemiology, health management systems, stochastic modeling, information systems, algorithms, engineering of economic risks, computer methods applied to the management business information systems, etc.

Researchers are studying themes that combine theory and applications, because the problems applied are sources of theoretical questions and the fundamental knowledge makes it possible to provide effective answers to practical questions. But there are also plenty of branches where it's just primordial.

Finally, the extreme rigor to which mathematics students are trained is appreciated in various private sectors (management, it, finance, organization, etc.) Mathematics continues to enrich itself for more than two thousand years of new results. But, between these extremes, there is a certain vagueness.

A computer variable and a function in the mathematical sense do not really have a relationship. To illustrate, we may additionally supply the fact that Haken and Appel surprised the community by offering evidence based on the use of the computer. On the other hand, mathematicians teaching in institutions without mathematics laboratories experience their isolation in a painful way.

Applications of this type are recruited young physicists, chemists (who have a good knowledge of materials), engineers, etc., who master the mathematical culture, scientific computing, methods of numerical analysis, etc. To illustrate, we may additionally supply the fact that the 80s saw the development of restrictions of first-order logic to clauses (conditional logic) as a systematic paradigm of non-deterministic programming, called logic programming, and popularized by the PROLOG language. Finally, the problem is much more complex than it seems.

On the other hand, to provide an example, we may also use mathematical and computer techniques as a development engineer. In a company, he is responsible for the development of medical imaging software. To illustrate, we may additionally supply the fact that mathematicians Wolfgang Haken and Kenneth Appel used a computer to demonstrate the four-color theorem.

Research in mathematics and computer science makes it possible to develop techniques, guaranteeing safety, more reliable than conventional solutions. So here we are without any common language or support.

The theory of finite-structure restricted models was born and has undergone significant growth in large part for the strong links it has with the theory of complexity or with areas of computer science such as constraints, database theory, games, auditing, etc. This will then be extended to the methods of undefinability (per game or combined methods). We will also be interested in the complexity of so-called "model-checking" problems and queries for some natural logic (on classes of particular structures, general, graphs of bounded degree, trees, etc.).

To illustrate, we may additionally supply the fact that if tech employees have never studied maths and when there was a documentation about a matrix for using a sensor, they would be confused. On the other hand, to provide an example, we may also use mathematical and computer techniques to determine a tour of a traveling salesman. It is difficult to determine a tour of a traveling salesman connecting the fifty capitals of the United States, while we now know, for this problem of exponential cost, exactly how to deal with such graphs.

To illustrate, we may additionally supply the fact that we may use mathematical and computer techniques for: complexity in communication, to compress, algorithms, streaming, to encrypt, to represent sounds, computational complexity, to represent images, etc. These behaviors vary according to multiple parameters to circumscribe.

An algorithm is a recipe that solves a problem in a systematic way. To illustrate, we may additionally supply the fact that in each datagram, the data is preceded by a header. These laboratories and their research programs, recognized nationally and internationally, are strongly supported. Its mission is often carried out in collaboration with the information systems department.

It is possible to propose a classic proof, which may, however, be very long, especially for a property which is not in the class NP. It is possible to show cause of the fact that a property is true with great probability. It is the role of interactive proofs and verifiable evidence with high probability that correspond to complex classes. If one looks for very fast proofs, one can introduce an approximation ε and decide if the structure satisfies the property or if it is ε-far from satisfying it, by using a distance between structures. If the structure of size S is very large, independent and independent proofs of proof may be desired only from time to time.

Lambda-calculus, aka λ-calculus, introduced by Church in the past 30 years, is at the frontier of computing and logic. Lambda-calculus is at the same time a system notation for intuitionistic demonstrations of the elimination of cuts, a formalism of representation of recursive functions, and the theoretical basis of functional programming languages. We will present the construction of concrete models and lambda-calculus systems, the categorical notions necessary for the definition of lambda-calculus models, the fundamental syntactic properties of pure lambda-calculus, etc.

Along the way, the necessary notions of classical, intuitionist, and linear logic are continued.

To illustrate, we may additionally supply the fact that there are professor-researcher or researcher positions. On the other hand, to provide an example, we may also use mathematical and computer techniques in a university (mathematics, computer science, mathematical logic, etc.) or a public research organization.

To illustrate, we may additionally supply mathematical and computer techniques for the research and development departments of industrial companies. The research and development departments of these companies are particularly demanding of students with a strong competence in logic, mathematics, and computer science, allowing them to supervise engineers working in the fields of program verification, software certification, etc.

It is on this ground that will develop what is, first of all, a set of technical know-how and give birth to computer science.

Let's first say that, in the face of Moore's Law, the doubling of accessible or storable information currently obeys a growth rate of a factor of two every 8 months or so. It would be desirable, for example, to teach computer science, algorithmics and programming by college, and to reserve mathematics and thrusts for those who want to specialize in this way. From the creation of video games to the development of web applications to the management and security of information flows, in all its forms: some projects that you do not have the burden.

So, if only to maintain consistent performance, new algorithms to be permanently designed. The compression algorithms are based, among other ingredients, on tree structures and on the Fourier transformation. Formal syntax studies, including variants of Lambek's categorical grammars, have been applied to proof theory formalisms such as non-commutative linear logic.

It's true that a person can take Java, take a list, type something and then receive the result, without even knowing what Java is. The advent of DVDs has not always been possible thanks to the development of codes based on the exploitation of the finite body structure and the teaching of the very large correction capacity – a scratch of the width of the body a hair can cause the loss of several thousand bits. You can find everything in the web domain and I have time to work with developers who have learned on the job.

To illustrate, we may additionally supply the fact that transmission layers are based on error correcting codes (finite fields), on the signal theory used in the design of modems and ADSL, and on the compression of binary data (performed by tree mathematics). By algorithm is meant a systematic method of calculation. The temptation may then exist for some leaders to perceive computing only the technological development needs of the current period.

On the other hand, to provide an example, we may also use mathematical and computer techniques for a weather map that we are going to load and show predictions based on various physical models based on the theory of partial differential equations, which are numerically processed by finely optimized mathematical algorithms (finite elements, meshes, discretization, domain decompositions, etc.).

And it is clear that being good in math is inevitable in the field of research. To illustrate, we may additionally supply the fact that an advertisement for a job offer from a bank looking for a developer to master the resolution of the heat equation (or Fourier); indeed, we find the same equation in the evaluation of certain options.

Indeed, to use their program, it is very complicated because it requires quite advanced knowledge (which may seem logical). The number and size of these configurations being commensurable with what the computers of the moment allow, an exhaustive examination of the special cases concludes the proof. Similarly, the finite fields are at the basis of many coding processes ensuring detection and correction of errors in the transmission or storage of data.

Mathematical and computer techniques is a formula in its own right, which can notably be used as a sub-formula. The satisfaction of controlled work is only much greater. These are used by computer scientists, mathematicians, and more generally scientists of all disciplines.

Algebraic languages, recognizable by stacked automata, are particularly important from the point of view of formal linguistics, and served as a framework for the formal grammars used to define the syntax of programming languages and other systematic notation systems. The computer programs thus become skeletons of the proof of their adequacy to formal specifications. Finally, loaded web pages are structured objects expressed in a description language (HTML) which also relies on syntactic and formal methods.

This domain is based on the theory of information found in statistics and financial analysis and goes through the formalization of the notion of source, which borrows in its mind as much from the theory of probability as the ergodic theory. On the other hand, to provide an example, we may also use mathematical and computer techniques to maintain efficiency on dynamically changing data.

Very quickly, it was recognized that mathematical logic was one of the main foundations of computer science. Ordering automatic calculators is neither simple nor obvious, and at first glance satisfactory methods prove futile in the world of computers. My opinion is that good basic mathematical training is useful to all.

Do you know that mathematics is at the base of many disciplines and current sciences? Substantial algorithmic progress has been made in areas related to systems of differential systems, polynomial equations, symbolic integration, series manipulation, and so on. This extremely fruitful point of view was to be developed bringing together logicians and computer scientists in search of the logical foundations of programming.

This research, originally motivated by heavy computational physics needs (e.g., disruptive development), now leads to a small number of large generalist systems, which are real factories of mathematical algorithmics based on millions of instructions. Higher-order unification studies provide elliptic processing, uniform modeling of anaphora search, and many other linguistic phenomena. The basic types include integers and real ones with arbitrary precision (only limited by the physical capacities of the machines), which allows, in particular, the experimentation in numerical analysis involving very unstable or chaotic phenomena.

On the other hand, this research makes it possible to design sophisticated programming environments, assisting a designer to build software that conforms to his specifications, or even synthesizing a correct certified program based on the proof of coherence of his specifications. In this section, we examine some of the contemporary trends in which there is a particular synergy between mathematicians and computer scientists. All this without losing sight of the fact that mathematics related to computer science are mathematics and can be studied for themselves.

And to understand the mathematical logic allows to better understand the reaction of our applications. In a word, you have to know how to think and reason. The extension of these notions to distributed and parallel computing is still a very active research avenue. These theories constitute the second basis of computing: as soon as it is necessary to define programming languages specific to the unambiguous expression of algorithms, to formalize the notion of algorithm, to check the coherence of languages and programs, etc., they will prove particularly valuable.

The calculations being carried out by the scientists themselves (or their followers), the notion of complexity is already very present: we know or we perceive that such a process is more efficient – converges faster or requires less manipulation – than another but the notion remains informal and subliminal. IT is both a science and a technology.

Whatever your current level of study, whether you like equations, numbers, management, math courses or to discover new things, mathematics and computer skills are for you. On the other hand, every calculator is from the operative point of view a finite system, acting on data that are also finite and plunged into countable sets as soon as we abstract from secondary limitations.

On the other hand, I also noticed that these people are hardly able to simplify things and to level themselves at the level of a layman. Then there is a more advanced training that depends on the orientation of the studies, although … The algorithmic problem would not become in these conditions somewhat obsolete or obsolete?

Other similar formalisms, like Milner's-calculus, were later invented to model the competing processes. The concepts resulting from combinatorics have led to the computerization of a rich set of data structures, among which, in the first place, the tree structure: word trees for dictionary organization and data compression, search trees for the structuring of one- or multi-dimensional data, trees terms in computer algebra, game trees (cf chess programs), etc. Thus, the field of speech processing, using mathematical methods of signal processing, and appropriate statistical modeling is in the process of providing human-machine interfaces using voice, usable by speakers without learning in restricted semantic domains.

For the major categories of problems cited, it is clear that a mathematical approach plays a crucial role in the formalization of problems and the constitution of a framework of thought. These considerations argue first and foremost for the importance of strong support for basic computer science research as a token of the future.

The generation is represented by the mathematical and computer system whose success has already been highlighted. Here again, the discovery of probabilistic primality tests which indicate without absolute certainty, but with a risk of infinitesimal error, that an integer is prime was a theoretical surprise, moreover of great utility in cryptography. Among the proliferation

are already strongly present in Greek science: Archimedes and Diophantine "calculate" the area under a parabola and the solutions of systems of equations in integers, while Euclid disengages the notion of an axiomatic system for elementary geometry, and Aristotle abstracts discourses for propositional logic.

38
CHAPTER

AXIOMS

Yet, currently, they do not necessarily lead to interdisciplinary work, neither students nor professors. Better yet, to find the ultimate scientific explanation of a phenomenon is no longer just to find the mathematical equation that explains it but rather the computer algorithm that simulates it. Gradually the logic is mathematized with a theory of models that uses the mathematical structures of the theory of sets and universal algebra (categories). The theory of the demonstration which is concerned with notations. The so-called binary logic means that a logical statement presents only two alternatives; it is either true or false, any other possibility being excluded.

On the other hand, to provide an example, we may also use and summarize the main algorithms and basic data structures of non-digital computing. We will explain in this chapter the main articulations of these currents that deeply interpenetrate mathematics and fundamental computing.

But unlike them, its destiny is not to be executed by a machine or even typed on a computer: it can be hastily traced on a corner of a tablecloth or a piece of greasy paper, to be transmitted in own hands to other project collaborators. On the other hand, to provide an example, we may also use and show the essential places in recent developments in mathematics, both statistics, probabilities and computer science.

These domains are based on the theory of information (found in statistics and financial analysis) and goes through the formalization of the notion of source, which borrows in its mind as much from the theory of probability as the theory, college students need to be aware of the role of mathematics in other disciplines and in all kinds of industries.

Without basic research there would indeed be no practical applications but do you really believe that all your students will invent, create, innovate also based advanced math? On the other hand, to provide an example, we may also use a construction of random sequences of a philosophical discussion on order and chance, as well as on the distinction between what is known before calculation and what can only be known after calculation. This frame now pervades the whole of computing; in turn, it gives rise to the emergence of new methods designed to circumvent the complexity barriers of the NP class: approximation techniques, probabilistic approaches, "parametric" refinements are examples.

I hope it sounds absurd to you. The research in combinatorial optimization is no longer the same-it is infinitely more structured and fruitful-after the work of their successors.

On the one hand, the typing of programming languages makes it possible to certify that a program accepted by the compiler (which includes a static analysis module verifying the adequacy of the program to the type system, possibly by synthesizing the type most general) cannot fail at runtime, nor can it corrupt data structures.

Historical research on algorithms and their achievements is not limited to recent times but extends to antiquity and non-European cultures. I then continued my studies with a training in imagery, and if I sometimes very much regretted not to have been sufficiently interested in the mathematical tools which had been presented to me in a very (too) theoretical way before, because I wasted time getting them into practice, but I still got my diploma without any worries, as well as all the others.

For college, we read in the report: to have computer science taught only by mathematics professors could lead to teaching only part of the subject, the one that comes closest to mathematics, which does not correspond to the aforementioned objectives. We understand this concern not to limit the teaching of computer science to only one of the components of this discipline, but, in any case in the current period when one must resort to professors already trained in pre-existing disciplines, must note that it is those of mathematics who have been the most numerous to propose to animate the specialty. The boundaries between these sectors are blurring and digital technologies are making these synergies possible.

39

CHAPTER

THE INCOMPLETENESS THEOREMS

To train those who wish to pursue scientific careers or to enable all young people to become enlightened citizens, mathematics professors are and will continue to be indispensable. This sector is indeed a big consumer of heavy digital calculations. This is, again, too restrictive a vision: many algorithms, to provide an example the recipe for food, operate on data that is not symbolic: flour, eggs, etc.

These skills exist in the training market. For that we need a language. The shift registers serve as memory support and as a tool for synchronizing signals in digital systems. Only one, because the half which one isolate immediately would have the size of the dictionary of the preceding example.

The algorithm progresses very quickly towards its solution: we will have read only headers before reaching the desired page. This methodology is gradually supplanting the traditional methods of operational research in some areas, and is therefore an important economic issue. The calculus is not just a notation for functional expressions.

In the positive case give his nature. The annex contains additional information on innovations. Finally, if we both rely on procedures generating sequences of numbers in order to elaborate their ideas on the nature of randomness, their reflections are exercised in two completely different contexts. Coupled skills are needed here: computer science, mathematics, chemistry, biology, statistics, etc. On the other hand, to provide an example, we may also use parallel computing with threads, processes and scheduling.

In fact, there is a large computer research community working in the field of big data and data mining. Math is first and foremost logic, it is elegance in the formulation of an equation that it is simple or terribly complex.

On the other hand, to provide an example, we may also use computing at the college and in the base of competencies: what proposals of interaction between mathematics and informatics? Their examination from the perspective of the hazard meets the requirements of the algorithm analysis described above. Today with hindsight I understand that they were right, my problem was at the level of my interpretation of the discipline.

The professor makes it clear to the students that the work that has been done is in fact the one that is done by the computer: one must transmit an image (by email to provide an example) to his/her comrade: the computer used transmits this image to another computer only using binary information (0 or 1: absence or presence of electricity).

The same applies to the divorce between science and letters. On the other hand, to provide an example, we may also use and build a tree of nested partitions associated with this relation. Thus, the algorithm that our cells use to synthesize a protein from a messenger RNA strand did not, as far as we know, have been written or verbalized, before being executed in the first cells, then passed from generation to generation. Remember, combinatorics counts among its origins several games or mathematical recreations.

Note that these same notions are marginal within academic mathematics, still dominant today. No, it's not even worth asking the question, it's so obvious. The sequence of actions to be performed for a robot, is formally identical. Computer science (reduce the place of the image, so code it). In fact, everything comes at the origin of confusion over the term computer. To illustrate, we may additionally supply a theory of models that uses the mathematical structures of set theory and universal algebra and categories.

After reception, the tape is removed and the paper copy of the packet is digitized and transformed into a transmittable electronic format. It is taught in computer science colleges in the same way as java, php, c, and SQL. This shows how slow and difficult the implementation of the program instructions of in the form of the study of situations from these disciplines including a modeling phase and an interpretation phase of the results. To illustrate, we may additionally supply that since the base 2 writing of a number can be very "long" and because of the simplicity of conversion between the bases 2 and 16, the computer hexadecimal system is very much used.

Multidisciplinary teams are therefore formed around this issue. Statistics and computer science, can also be applied to chemistry, biology, etc. To illustrate, we may additionally supply and therefore perform the Euclidean division of a natural integer I by a nonzero natural integer B is to determine the unique integers Q (called quotient) and R (called remainder) such that: I = BQ + R and $0 \leq R < B$.

But these contents are not so important: the vast majority of students will forget the details of what they learned in college, and it does not matter: there is not a theorem that is indispensable in life daily. In mathematics, on the other hand, apart from probability, most of what we learn dates back to before the 19th century. On the other hand, to provide an example, we may also use an impact of pedagogy specific to computer science, and this report recommends a large place for projects.

We should also read the presentation of general principles to guide the teaching of computer science. These ancient traditions developed fully in the Age of Enlightenment, and it accompanied the mathematics, and mathematical logic revolutions, when science played an extremely important role. So much so that computer science introduces a very general notion of language, of which natural languages, programming languages, but also mathematical language or the languages of logic, are special cases.

After that the need for computer math seems obvious to me! On the other hand, there are a lot of branches in maths, and all that concerns the lessons related to discrete mathematics to provide an example do not seem to me useless to succeed his course.

To illustrate, we may additionally supply computer devices that are autonomous and must, to be effective, adapt in real time to their environment (GPS, autopilot, etc.). This division of labor creates an exponentially growing body of knowledge, the assimilation of which by society is a very difficult problem. On the other hand, to provide an example, we may also use new organization data structures such as databases. It can improve relationships between professors and students, and bring diversity to the activities and skills developed.

We find, in a paragraph evoking the components, algorithm, language, information and machine, in the current program of the specialty. To illustrate, we may additionally supply the fact that my graduate university degree came from a computer training oriented development. The best current data structures are often therefore based on a clever use of tree structure, a natural derivative of the notion of graph. Integration of new technologies into teaching what is the specificity of the links between mathematics and computer science? The assisted demonstration of theorems has also developed the extension of the search for solutions to diophantine or polynomial equations to various and varied algebraic structures, allowing the calculation of pattern search and filtering or solving equations between terms and unification.

On the other hand, to provide an example, we may also use mathematical and computer techniques for k-types, results of definability, theorems, forecasts, locality, optimization, games, decision support methods, data analysis, etc. On the other hand, to provide an example, we may also use mathematical and computer techniques which allows, in particular, the experimentation in numerical analysis involving very unstable or chaotic phenomena.

Biology and chemistry make it possible to understand the context in which statistical and computer tools should apply. But it's hard to change habits. To illustrate, we may additionally supply a notion defined to model the human language. This type of work can arouse students' curiosity and make them perceive new reasons for disciplinary content by allowing other approaches. To illustrate, we may additionally supply, if we are specialized in networks and software engineering, we will therefore develop tools related to digital processing and the design of routers, networks, and communication protocols.

Mathematics has always had a strong connection to their teaching. To illustrate, we may additionally supply and identify how it works and how it breaks down. To illustrate, we may additionally supply mathematical and computer techniques for counting, combinatorics, solutions of recurrence equations, permutations, recurrence relations, arrangements, etc.

And we should participate in its beginning at a national council of mathematical and computer techniques for universities. But there are other areas that overlap the two disciplines on a border each day more important. This is as early as elementary college. On the other hand, should therefore be setting up multi-disciplinary modes including computer science and touching mathematics. curriculum reorganization, especially in hourly volumes of the different disciplines.

The basic types include integers and real ones with arbitrary precision only limited by the physical capacities of the machines. To illustrate, we may additionally supply mathematical and computer techniques for mathematical tools, computer tools, statistical tools, probability theory, etc. A widely held but false idea is that an algorithm is necessarily deterministic: it leaves no room for hazard.

On the other hand, to provide an example, we may also use mathematical and computer techniques for analysis and modeling and equations of phenomena that have been noted for their importance. At the dawn of the third millennium, the question arises as to whether this period of pioneers of computer science based on the mathematical sciences is or is not in the terminal phase. It is taught even in management colleges.

The debate underlies a problem partly epistemological, made difficult by the lack of knowledge and investment in the interface between maths and informatics of too many mathematicians. However, many professors who have participated in mathematical and computer techniques are satisfied with it.

Even if the structural properties of these algebras are relatively poor, the associated algorithmic is de facto one of the key points of the logical verification of computation circuits by binary decision diagrams or constraint systems. At the border of the notions of language and machine appears the notion of machine language. But if computer science leads us to see today in discussion an interesting text for the history of mathematics, we must nevertheless use the concepts of modern computer science with caution when interpreting historical texts.

The concept of information and a concept of language appeared in computer science, mainly because, to communicate algorithms to machines, we have to use a formal language. To illustrate, we may additionally supply non-deterministic algorithms. No need to destroy hardware for crash tests, the engineer does everything on a computer.

But if the Internet was first designed for computers, it does not mean that it only makes sense on computers. Maths are above all the most fabulous language of description that we have invented, obviously that everything can be described with. On the other hand, to provide an example, we may also use codings proposed by pupils that revolve around parameters: the color of a token compared to a black and white color of an image.

On the other hand, the main opportunities for a mixed training in Mathematics and Computer Science are very varied. It is very likely that, whatever the fate of the various recommendations of this report, the teaching of mathematics is, more probably than that of any other discipline, jostled in the years to come by that of computer science. One can then enumerate the operations associated with each class of objects: a member can be registered, receive a mail, deleted, etc.; a work can be recorded or deleted; a copy may be registered or deleted; etc. In particular, it makes it possible to match a computer program to the formal proof of a mathematical statement, which carries out this statement as a specification.

On the other hand, to provide an example, we may also use mathematical and computer techniques as a modeling engineer. And write them a necessary step to communicate them to a configurable machine. A young engineer, recruited to provide an example, will probably make a career there, if he gives satisfaction. A little to assess the risk of sharp decline in stock prices, to provide an example. It turns out that many mechanical actions, all probably, lend themselves to such examples.

I would confess you that I read diagonally mathematical and computer techniques articles, I answered especially to the computer science, mathematics, and mathematical logic. All the other scientific disciplines have their own object, in the nature or in society. For this, we identify the main types of objects, their attributes, the operations they can undergo and their interactions.

On the other hand, to provide an example, we may also use experiments that have been made on learning geometry in the middle course in relation to physical situations (mirrors, compasses, sights used to introduce the notion of angle, to provide some examples).

Their specificity, so to speak, is that they are not specific to any area of nature or society. The sense is that of method to perform the operations, but it extends to any method of calculation. Mathematical and computer sciences are crucial and everywhere present behind this now banal act of everyday life.

Nowadays, many mathematicians use experimentation whose possibilities are increased tenfold thanks to the formal calculation systems described below. We pointed out the successful development achieved in the mathematics programs of the scientific preparatory classes. On the other hand, to provide an example, we may also use mathematical and computer techniques for computing, management, statistics, etc.

These groups must be able to provide a common explanation, which allows a misunderstanding of misunderstandings and a common work on language. However, it is important to note that the first universal programming languages, lambda-calculus, the language of Turing machines, predated by a few years the construction of the first universal machines, the first computers.

Finally, let us detail the possible interactions between classes of objects: a member can book an album, an album can be embodied by a copy, a copy can be borrowed by a member, etc. To illustrate, we may additionally supply computer science, mathematics, and mathematical logic for a collection of information on historical bases, stochastic tools, probabilities, statistical tools, etc.

The central part of UML, the class diagrams, is to draw the project guidelines. To illustrate, we may additionally supply computer science, mathematics, and mathematical logic for statistical analysis of experimental results. But mathematical calculations form only a part of computer applications and they are likely to interest only those that mathematics are already interested in.

What place to give to applications? To appropriate a concept, it is not enough to read the definition, you must have time to handle it by doing exercises. But if you want to do development in R & D mode, it's something else. To illustrate, we may additionally supply a machine that performs a particular calculation: it raises the number I squared.

Combined skills (marketing techniques-data analysis-mastery of software or data mining techniques such as data mining are used in marketing departments of insurance banks, industrial companies or services, etc. To illustrate, we may additionally supply the part already taken by learned societies and professional associations.

This is an opportunity to focus for a few minutes on a particular point. Similarly, by grouping in a sequence of 0 and 1, the symbols by two and replacing the group 00 by a, the group 01 by b, the group 10 by c and the group 11 by d, we obtain a sequence two times shorter but in a four-letter alphabet.

Thinking is the nobility of the mind, which is difficult to explain. We can finally ask students about the number of tokens used and try to reduce this amount if it seems possible. There are many companies that hire young people with a solid mathematical culture combined with a good knowledge of computers in a logic of mastery of software engineering (in this sector, computing power is sought). If you find one, even if you are unsure, send me an email.

They call on high-level specialists who have accumulated a very large experience in their profession. A specificity of mathematical and computer techniques curriculums is therefore to offer end-of-year internships oriented towards research. We alert them from the beginning: for us, it is first to learn theorems, models. Considerations of formal linguistics and logic would thus mark fundamental computing.

The collected data (clinical, socio-demographic, biological, etc.) are analyzed and interpreted thanks to the cross-skills of statisticians, doctors, computer scientists, biologists, etc., and come to nourish the files mentioned above. These theories constitute the basis of computing: as soon as it is necessary to define programming languages specific to the unambiguous expression of algorithms, to formalize the notion of algorithm, to check the coherence of languages and programs, etc., they will prove particularly valuable. The aim is to offer mathematics students a solid openness to mathematical and computer techniques, accompanied by an orientation towards privileged mathematical and computer domains.

Often we do not even have to touch it because there are ready-made functions that do the math for us. Well, it's still maths. Identifying all the functions that calculate the sorted form of a list prevents you from looking at the intentional properties of their algorithms, such as the complexity of the calculation. To illustrate, we may additionally supply, discover, and popularize the importance of probabilistic approaches in the design of algorithms.

A balance sheet must be drawn. They make it possible to study the behavior of consumers with a view to establishing typologies to adapt them to the sold service products, and to analyze multiple data coming from the markets. As soon as a problem is solved, it opens the field to new problems.

The students therefore understand that one can represent a complex information (image) with very simple objects (continuation of two-tone tokens). This type of problem is very common in computer science, and more precisely in algorithmics, the science that seeks to optimize calculation methods.

I must say that I really understood the objects only after thinking about them in relation to vector spaces. We will also give the new matrix. Formal syntax studies, including variants of Lambek's categorical grammars, have therefore joined the formalism of proof theory such as non-commutative linear logic.

Many graduates believe that computer science is very concrete, it is doing and making websites. Their field of action is endless: health, climate, marketing studies, etc. The sectors covered in this have many strengths: varied opportunities, attractive salaries, real opportunities for change and places to take.

I pretty much agree with this if we talk about pure mathematics of studying complex functions, signal processing for those who do not appreciate not maths, solving differential equations, and so on and better. Mathematical algorithms always operate on mathematical objects: equations, matrices, polynomials, numbers, etc.

An encouraging regional context and the scientific excellence of the laboratories convinced therefore the institutes for research in mathematical and computer techniques and the institutes for research in computer science, whose vocation is to undertake fundamental and applied research in new information technologies, to create research units. The algorithms discussed here therefore are based on the fine analysis of semi-classical probabilistic phenomena, on the already discussed randomization technique, as well as on interesting methods of complex asymptotic analysis.

To illustrate, we may additionally supply a class that is represented by a frame, in which we note the name of the class, the list of attributes, then, under a horizontal line, all the operations that the objects of this class can do. On the other hand, to provide an example, we may also use a very complex system that incorporates many physical and chemical phenomena. Mathematics and computer science should therefore provide a background of a rather broad mathematical culture with an emphasis on applied algebra.

The number of installations in the world of computer algebra systems can be estimated at a few million.

Fuzzy equivalence relations and closure is obtained by successive composition with the initial relation. There is, in mathematics, therefore, a process of perpetual renovation. To illustrate, we may additionally supply IP and provide delivery attempts that will be repeated delivery attempts until the delivery fails.

These procedures quickly became very popular. Moreover, students sometimes seem sure of them. To illustrate, we may additionally supply a test of the generalizable character of the created models. To answer these questions we should also need to invite to these hypothetical conferences the advice of several great scientific colleagues.

On the other hand, to provide an example, we may also use a factorial, that is very often used to illustrate the principle of recursion, just like the arithmetic sequences and whatnot. What are the interrelationships between the teaching of computer science, the use of computer tools in mathematics and digital tools for mathematics teaching? Many training courses of this type are available on the training market. In the same way, the grammar of a natural language never seems completely known, as evidenced by the difficulty of doing orthographic or automatic translation programs.

At the border of the notions of an algorithm and language appears the notion of program. The goal is to evacuate the thought of the calculation, in order to make it executable by a digital machine and a computer. On the other hand, to provide an example, we may also use mathematical and computer techniques for informatics, statistics, mastery of "big" software, etc.

I agree with the fact that we give too much importance to mathematics, but not only in computer science. To illustrate, we may additionally supply therefore develop numerical simulation methods that show a continuity between mathematics and computer science.

On the other hand, to provide an example, we may also use the period from 1940s to 1950s, that was one in which the merger between scientific calculators and management calculators gradually took place, a field in which the powerful IBM company flourished. Thus in the primary is recommended the first approach of algorithms; why not encourage us to make the link with the elementary calculation algorithms?

In more than fifty years of life, Moore's law has not been denied and physicists believe that its applicability will extend until reaching the limits imposed by the atomic structure of matter and quantum phenomena. The development of the field is therefore carried out by teams with mixed skills where it is sometimes difficult to distinguish the part of each other. To illustrate, we may additionally supply and offer invitations to students to show things to know before it's a fact.

Whatever his specialty, a computer scientist will always be called to repair his brother-in-law's computer. It's just an empty word. Yes, I also dump on some points in math and other related subjects, but it allowed me to push me even more. Recall that logic itself, initially a preoccupation of philosophers with the notion of truth, gradually came to be mathematized.

On the other hand, to provide an example, we may also use the theory of demonstration or meta-mathematics that is concerned with evidence, notations, and more generally underlying combinatorial structures. To illustrate, we may additionally supply mathematical and computer techniques for numerical analysis, mathematical modeling, simulations, statistical modeling, etc. Therefore, to really access computer science, you must have a university degree in computer science or mathematics or mathematics and computer science.

It is urgent not to wait anymore. I therefore call for the integration of computer science education at all levels of education and in all sectors. The mathematics strategy should therefore be in which computer science plays a major role in teaching, an example below is taken from a hypothetical dossier presenting this strategy and professor training for a creation of a computer option in mathematics. It is difficult to answer this question. But it is striking that here no more mention is sometimes made of possible coordination with mathematics, while openings are made to the functioning of enterprises, to technologies, to the human sciences, to laboratory methods, etc.

In an industrial environment, many physicists and chemists are hired to practice modeling and simulation in order to predict the behavior of industrial processes, products and processes materials. So, if only to maintain consistent performance, new algorithms must be continually designed, as we observe. To predict an approximate value of a result is very interesting. We also saw the emergence of a theory of complexity, an area inspired by an abstraction of the complexity of algorithms and rooted in the theory of computability discussed below.

It is a necessity for students to truly learn computer science. Finally, new logic and architectures of formal proofs have been applied both to the proof of major mathematical theorems and to the recent formal verification of compilers, circuits, and operating systems. These recent cases illustrate well one of our theses, namely that the relations of dialectical exchange between mathematics and computer science are far from being completed.

On the other hand, to provide an example, we may also use mathematical and computer techniques for survey techniques, probabilistic tools (because you have to master randomness), statistical tools, data analysis, studies, etc. To illustrate, we may additionally supply an analysis of data from social security sheets. We should also point out that mathematical operations were often used to learn programming.

I personally think that mathematics is used to build the tools that computer scientists use. With hindsight, we realize that to make a translation of a text in Latin is not very far in the reasoning of the writing of an algorithm corresponding to a given need. To illustrate, we may additionally supply a logistical problem: how to put storage points and warehouses at strategic points over a defined geographical area to better control storage times and costs and indicate all the partitions that can be deduced from these relations. We should also emphasize their magnitude and urgency.

To illustrate, we may additionally supply mathematical and computer techniques for cancer research and its foreseeable evolutions. The fields of application are many: software navigation in geographical data (Google earth), automotive remote control (gps), videoconferencing, etc.

I was less fanatic about geometry in the space that is less suitable for the demonstration. It is important to do this regularly, at college, using as many applications as possible. Moreover, in a numerical calculation, the attention paid to orders of magnitude. To provide an example, deciding whether or not a sentence is well formed in a language requires using an algorithm that is independent of the fact that this language is a specification, programming, query language, etc.

It is an illusion to think that students only know what they are taught at college. But overall understanding is more important than learning by heart. To be successful, having a BA/BS, with math skills, is strongly advised. Moreover, a student would feel very strongly when in a final year, a math professor gives them an application for post-baccalaureate studies with the mention: does not have the required skills to follow this formation.

To illustrate, we may additionally supply mathematical and computer techniques for functional analysis, algorithms, complex equation solving, scientific computation, etc. On the other hand, to provide an example, we may also use UML as a programming language on paper, which is executed only by human beings. To illustrate, we may additionally supply mathematical and computer techniques for work as a: research engineer, telecommunications engineer, epidemiologist, laboratory engineer, development engineer, test engineer, clinical research associate, medical imaging engineer, qualification engineer, biostatistician, etc.

Whenever possible, they will be linked to the teaching of other disciplines: technology, life sciences, geography, earth sciences, etc. The discussion is a philosophical and epistemological issue. And if one wonders what they are intrinsically?

To illustrate, we may additionally supply mechanical machines, before we have the technical knowledge to build the most common tools today: computers. In computer science, math is useless for a large number of students. On the other hand, apart from references to open and massive online courses, we give little indication of its content.

Because learning computer science, it is not learning to click or write with a word processor, it is above all learn to program. It is indeed surprising to discover that very classic areas of mathematics, developed most often without immediate application, such as combinatorial analysis and asymptotic analysis (real or complex), have to do with such an artificial world computers and their programs. To know if you are made for a computer science degree, do not hesitate to take a test.

Many innovative ideas have been generated by such motivations, and their interest has been understood later. To illustrate, we may additionally supply the idea that database specialists will immediately know how to convert a diagram into a database schema, and object-oriented programming specialists how to build the program and the interface to use it. These disciplines combine in the reasoning applied to finance and insurance.

I'm not saying that we should not teach mathematics in computer science, but that they should be less dominant and a good level in this subject should not be a prerequisite, or if it is, not be so important. Technology is the tip of the iceberg, the one facing the general public.

To illustrate, we may additionally supply mathematical and computer techniques for a theory of optimal control and search for a trajectory that makes the best use of the gravitational attractions of the various stars in order to minimize the fuel used by the probe. This does not prevent the mathematician at the work of being motivated mainly by the beauty of a problem (whether it is elementary or not), the desire to better grasp a property, or pure curiosity.

To illustrate, we may additionally supply a meaning that is that of a method to perform the four basic arithmetic operations (formalized by Al Khwarizmi in the ninth century), but it extends rapidly to any method of calculation. In college, we know that programs in mathematics have recently evolved through the introduction of logical and algorithmic chapters, not forwards to computers; we therefore regret not to see how to accompany this evolution. We cannot compare the case of computer science and that of probabilities.

I'm the same as you on this point: no math but good algo. To illustrate, we may additionally supply mathematical and computer techniques and determine the local extremums of a function and specify their nature. On the other hand, to provide an example, we may also use mathematical and computer techniques for a mastery of mathematical tools (Statistics-probability for use in the uncertain future), and Computer Science (representing the information of a mathematical model), as well as a solid culture in financial and economic techniques using mathematical tools.

To illustrate, we may additionally supply mathematical and computer techniques for the field of advanced specialties, web applications, project manager assistants, computer systems and software, professional licenses, digital images developer, etc. On the other hand, to provide an example, we may also use dynamic laboratories that are structured, in particular, with a high scientific output.

It is also interesting to note that a new generation of systems with better characteristics (security, key length, etc.) are in development and that many of them rely on the algebraic geometry of the curves and manifolds on the finite fields. Let's talk more specifically about mathematics education. To illustrate, we may additionally supply and write in bases 2, 10 and 16 and they are also called the binary, decimal and hexadecimal, respectively. In base 2 or 16, the operations are performed in the same way as in base 10.

Should we rethink mathematics by computer or is it the computer that allows us to rediscover certain nooks and crannies of mathematical discourse? On the other hand, to provide an example, we may also use mathematical and computer techniques at banks, actuaries, economics, mathematics, computing, insurance, etc. You may wonder, "Should mathematical and computer techniques be modulated according to origin of these professors?"

Adequate time is also essential for students. By algorithm is meant a systematic method of calculation. The procedure is perfectly suited to paper but would waste the resources of a computer. It is a question of characterizing the main parameters of large combinatorial structures, when this one obeys a determined random model. The answer will be precisely justified.

The breakthroughs of Rabin and Knuth, relayed by a large part of the computer scientist community, have changed the way in which algorithmic research is conducted, offering clear benchmarks in what would otherwise be a jungle of techniques. They are constantly seen to be outside the field that gave birth to them.

A major project is therefore, alongside the development of an appropriate initial training, that of the training of professors already in practice. To illustrate, we may additionally supply the forces that should be mobilized for this purpose. With IT, random simulations can be performed to assist in decision-making in many cases, such as: quality defect studies, hazard control in manufacturing lines, dependability studies, cell phone call forecasts to optimize the network's deployment, studies in order to minimize breakdowns, the search for the best price policy to be implemented by the airplane companies to ensure plane filling, financial risk assessment, survey institutes, market risk for banks, etc.

On the other hand, to provide an example, we may also use mathematical and computer techniques as an: information systems architect, econometrics, computer auditor, experts in the field, Internet technology, decision analyst, multimedia technology, controller, insurance engineer, etc. To illustrate, we may additionally supply an example: if we measure the time needed for the ball to travel a distance D, we obtain a machine that calculates the square root of the number D.

Math is everywhere around us. The number 23, written in base 10, is written 10111 in base 2 and 17 in base 16, but one cannot write 23 = 10111 = 17. Master the principles of numeration essential to the languages of low level. What place in teaching?

The link to physics deserves special attention. On the contrary, many algorithms need randomness to work. To illustrate, we may additionally supply an example (10) 2 + (100101) 2 = (100111) 2 (in base 10, this gives 2 + 37= 39).

I therefore have great difficulty understanding the importance given to mathematics in the context of computer learning. It is developed in modes of learning, roughly corresponding to elementary education (kindergarten and primary), college and college: acquisition of autonomy, discovery, deepening concepts, etc.

As for mathematical and computer techniques work, it takes form according to an entirely different materiality, that of paper and ink. On the other hand, to provide an example, we may also use IP addresses are used to identify the host of a network. There is no multi-diffusion, but thunderstorms can cause data loss. Of course, doing mathematics, contrary to what many journalists think, is not doing calculations.

It is not essential that they have been previously exposed in the mathematics course. The same physical system, equipped from another protocol, performs another calculation. New applications of combinatorial analysis and complex analysis (in a style that extends that of pioneers of the early twentieth century) are perhaps more surprising. The course of the session is the same, but students are questioned.

We will justify the results that will have to be given according to the values. A view of the contemporary movement of the mathematical sciences could be valuable. The relations between these different concepts are also

mathematical models and their use. We think here of mathematics professors, but the proposal does not would not be absurd for history professors

I have always been passionate about computers. On the other hand, sometimes my friends do not seem very convinced by my arguments, because they think that the manipulation of variables and their evolution over time during the execution of a program are typical of mathematics. To illustrate, we may additionally supply the fact that we may use mathematical and computer techniques for fractals, chaos theory, the theory of automata, numerical analysis, cellular analysis, etc.

Until the seventeenth century, there is often coincidence between mathematics and actual calculation. We will therefore be recruited for broad mathematical skills with skills in signal processing, celestial mechanics, fluid mechanics, software engineering, etc. On the other hand, to provide an example, we may also use mathematical and computer techniques for economic, statistical, financial, and studies relating to insurance services products.

40
CHAPTER

CONCEPTUAL PERSPECTIVES

On the other hand, to provide an example, we may also use computer science (for computer simulation, the latter offering the computing power required); mathematics (numerical analysis, scientific computing); and mechanics of solids. However, water has flowed under the bridges since that time and if the computer is still used today massively to perform complex scientific calculations, a good mastery of mathematics are far from a prerequisite to be programmed. Finally, the notions of language and information are related because they are two instances of the same idea: to symbolically express algorithms, for one, data, for the other.

Mathematicians are recruited to implement these advanced technologies. But not much about the essential, the functioning of the system itself and that of networks. The calculation has a sporting side that can appeal to children, sometimes developed phenomenally, to provide an example. We must think about the future, the inevitable evolutions of the programs and thus build step by step a solid mathematics and computer science discipline that can, when the time comes, the experiments made, stabilize. No, we do not use an external service to calculate an internal data.

On the other hand, to provide an example, we may also use computer science, mathematics, and mathematical logic for writing a matrix associated with the quadratic form F, and write F (x) in the form of an algebraic sum of squares of independent linear forms. These two remarks lead to a question that is not yet solved today: when can one say that two algorithms are identical? To illustrate, we may additionally supply a probability theory and has its roots in games of chance.

Finally, what place should computers be given in the history of mathematics? The notion of the machine finally appears in the renewal of the scientific instrumentation. On the other hand, to provide an example, we may also use and define in a set of variable "x" equations a multitude of physical, chemical or biological phenomena that evolve in space and time.

Finally, the path taken by the datagram can easily be traced: traces are automatically generated, and can often be found on logs and cables. Let us mention here that formal calculus can be seen as an extreme version (going as far as computerization) of the constructive current in mathematics, where abstract proofs of existence are replaced by explicit constructs. When they arrive at the preparatory class, they can speak, draw and count.

A mathematical result always has a story and changes meaning in the course of this story. It is, in the same way, a necessary training. They also suggest that the best way to train future generations of high-performance computer scientists is to provide them with training adapted to the objects, concepts, and methods of computer science, but above all integrated into a standard and generalist quality scientific training, and including, in particular, solid mathematical bases. Many in the audiovisual sector (cinema, television, multimedia, etc.).

It is no longer taught the same geography, the physical geography has largely passed in the Sciences and the discipline has moved closer to the social and environmental sciences. The first deals with citizen mathematics, the definition of which is not clear. By the way, from another university, but on the same platform, to provide an example, an individual may be taking a very interesting course. If several unified frameworks have thus brought In the quantitative analysis of computer algorithms and systems, it should be noted that large open questions remain in theory of the intrinsic complexity of problems.

Thus, a historical source of combinatorial analysis is a very theoretical need to understand the formal calculations underlying mathematical analysis. The theory of categories and mathematical logic seemed at the beginning to lie at the edge of the philosophy of science, whereas the theory of numbers had, in the eyes of the mathematicians. As to whether this justifies that we learn them from the college, it is to discuss again: they are delicate concepts, which require a great fingering to be transmitted profitably.

To illustrate, we may additionally supply a problem is that computer scientists do not know all the details of library administration and librarians are not computer specialists. The results will be justified without necessarily specifying eigenvectors. But they are experienced and exceptional professors by their culture and dedication.

I will report on it in a future chapter. The first step is therefore to derive a uniform functional notation to note the algorithm that, at the formal variable I, matches the value of an expression E containing occurrences of I. On the other hand, to provide an example, we may also use the common base of knowledge and skills, in the course of consultation, in which one can read in particular.

A general view of the science movement should be indispensable. The training in question disappeared with the reform, but I suppose that there are still today equivalents. On the other hand, to provide an example, we may also use computer science, mathematics, and mathematical logic to adjust land management accordingly to mitigate the negative effects of climate change.

Official instructions in recent years not to dwell on learning to calculate, to give up technical exercises, and professors complain about the very low level of their students in calculus, including scientific students, where many students are unaware, to provide an example, of the use of basic ideas. The realization of an IT project, to provide an example the IT portal of a library, requires the participation of many actors.

How should mathematics be adapted to be implemented on a computer? The results must be justified. The programs, which highlight possible reconciliations, very rarely mention a relationship with mathematics: to provide an example, the very interesting year program in the sciences of life and the earth (food, living world, environment), mentions the links possible with the programs of geography, technology, and civic education, forgetting mathematics. On the other hand, to provide an example, we may also use mathematical and computer techniques for chemistry, mathematics, physics, computing, industrial research, etc.

At the crossroads of these disciplines, we see that teams are formed and provide mutual insights against a backdrop of highly complementary skills. On the other hand, to provide an example, we may also use an interview devoted to the teaching of mathematics and study the nature of the computer science, mathematics, and mathematical logic. This is why their introduction is accompanied by a decline in the supply of mathematics education in most colleges and has been poorly received by professors.

What contributions of computer science for learning mathematics and vice versa? This is where the computer science, mathematics, and mathematical logic must be justified in my opinion. In terms of the method, it is also closer to the natural sciences because it relies on material objects – computers – to implement abstract algorithms.

To discover this notion, it is explained to us what an algorithm is: an obvious solution. On the other hand, to provide an example, we may also use general, technical and vocational colleges. But those who succeed are those where professors are recognized and have a high social status. Mathematics is often used in other disciplines to express laws in a rigorous and objective way, or to provide computational tools.

By discrete maths I mean optimization, automata, theory of languages, theory of graphs, etc. It is then an element for the mathematics professor, who can integrate into a structured teaching. Companies in this sector invest heavily in research and development (up to 15% of their turnover).

colleges and universities are not intended to train you for a job. Here again, it is not our intention to take a stand on the establishment of a specific body of computer professors, with the creation of an aggregation; this perspective is in line with the assertion of the autonomy of informatics as a new college discipline, amply argued in the computer science, mathematics, and mathematical logic.

The hand-to-hand formula, which renews the students' approach to observing and experimenting with natural objects, should be accompanied by the implementation of relevant mathematical concepts and methods. Finally, and this probably explains the remarks above, the grammar of a formal language is often much simpler than that of a natural language.

Some of the results are immediately valorized by industrial applications. To provide an example, the way in which one learns a foreign language, let alone a dead language, is sometimes closer to the way one learns a formal language than to the way one learns one's mother tongue. For complete information, see the rules and scholarship plans and the student guides of the different faculties of science available at the different colleges' section of mathematics.

The answer will be precisely justified. These differences are not only at the mathematical level, we do not yet know the theory of a more elaborate notion of randomness; they are also and especially in the concrete and material experience of the realization of mathematical calculations as well as in the historical context of philosophical debates on the notion of order and chance. On the other hand, to provide an example, we may also use the equations retained reflect relevant laws identified.

To illustrate, we may additionally supply a realization of this work in class. It is found in the detail of the sessions. I see main goals of the math course: a number one goal is to learn to follow logically. On the other hand, to provide an example, we may also use computer science, mathematics, and mathematical logic for adding two numbers, and we obtain the third term of the sequence by keeping only the number of units (modulo 10, we would say today).

To illustrate, we may additionally supply statistics and informatics and a mastery of software. The most classical mathematics, especially those inspired by the ancient needs of physics, are continuous. This field has been particularly active during the last decades, particularly with linear logic, the calculation of constructions, and more generally the work carried out at the institutes of mathematics. Thus, the algorithms have become programs. The goal of research is not to accumulate results, but to progress in the knowledge and understanding of mathematical beings.

The answer must be justified. There is no miracle, but things have a rational explanation. Every time we build a study program, we are witnessing Homeric battles. We must track down the differences to understand, measure, and appreciate the temporal distance. The construction of universal machines, however, has not completed the history of machine construction, since many improvements can still be made. These tools pose difficult problems of representation and quality of approximation of the quantities handled.

This activity allows, through the manipulation of objects, to understand the concepts of binary coding of information and thus to develop a concrete knowledge of an abstract notion. Banks have been using mathematicians for decades, while this recruitment trend is more recent in insurance. The tone was given by the conclusions of the interventions of biologists, and we need students conceptually trained in these mental operations that only give mathematics. On the other hand, to provide an example, we may also use mathematical and computer techniques studies, to provide an example, for the launch and the maintenance in orbit of a satellite by the optimal control theory and celestial mechanics, the optimal trajectory of a probe, etc.

Let's take this half between two hands, and open it again in the middle. At all levels, mathematical concepts will be used by professors in other disciplines. Scholarly and professional associations have already done a lot of work in this direction. The browser itself is a complex program whose construction has involved programming languages and interaction models benefiting from the syntactic and logical theories amply discussed above.

On the other hand, to provide an example, we may also use computer science, mathematics, and mathematical logic for philosophicals concern since antiquity, by its relationships with languages, and as a rational investigation of the notion of truth. In sum, computer science invites us to see some chapters of the history of mathematics from a new perspective, but we must not be seduced by the appearances of the same history.

On the other hand, to provide an example, we may also use computer science, mathematics, and mathematical logic and a person connects on the Web and obtains the weather forecast map of his region. This total absence of allusion to our discipline, whereas upstream had been noted the intrinsic character of the connection of computer science, in terms of the objects it studies, with mathematics, seems to us most regrettable. To illustrate, we may additionally supply jobs for manufacturers continue to gain market share and the sector is investing heavily in both capital and skilled jobs for research and development.

On the other hand, to provide an example, we may also use computer science, mathematics, and mathematical logic and give an example of two matrices having the same eigenvalues but which are not alike. There is therefore a certain danger in seeing mathematics develop for a particular specialty, to the detriment of what mathematics can bring as an element of common culture. I told myself, nothing to do with all these.

To illustrate, we may additionally supply in IPv4, still widely used in 2017, these addresses, known as IP addresses, are numbered from 0.0.0.0 to 255.255.255.255, thus passing through addresses as expressive and easy to remember as 192.168.0.1. This research gradually joined those led by logicians in the theory of demonstration, with the emergence of a fine combinatorics of proof structures (geometry of the interaction, linear logic, demonstration networks, etc.). Scientists and literary people came together to make algorithms.

Close collaboration is practiced between computer scientists, biologists, and mathematicians. On the other hand, to provide an example, we may also use computer science, mathematics, and mathematical logic for parallel and specialized machines. These logical methods are complemented by statistical methods, usable on a large scale now that large computer corpus is available (Web, newspaper archives, etc.).

The finite automata are a simple and rich computation mechanism: they correspond to a device that interacts with its environment according to a memory and finite rules. To illustrate, we may additionally supply an advantage of mathematics from elsewhere. This recommendation is not obvious. The number and size of these configurations being commensurable with what the computers of the moment allow, an exhaustive examination of the special cases concludes the proof.

The IP protocol is strictly adhered to: The IP packet is sent, in hexadecimal, the bytes separated. A testimony is the exceptional sessions held on this theme with the participation of mathematicians, computer scientists, biologists, physicists, economists, mechanics, and association leaders.

In short, as I always say to my students, mathematics is only for those who do the computer, not those who use it. During my studies I had continually study in mathematics courses and in algorithmic courses. On the other hand, to provide an example, we may also use mathematical and computer techniques to send probes and satellites into orbit at the lowest cost with the greatest possible precision in the aerospace sector.

Applied mathematicians and computer scientists collaborate in the development of numerical and computational tools used in a considerable number of scientific and industrial applications. Do we have good professors? Gradually, this problem has been enriched by a paradigm of constraint programming, aggregating linear programming methods, programming in integers, and more generally solving various algebraic equations.

On the other hand, to provide an example, we may also use computer science, mathematics, and mathematical logic and provide a project and case study on the implementations of an algorithm. To provide an example, when we write $ax^2 + bx = 0$ the "variable" x denotes the solutions to this equation, and so x take zero, one or two values at a time. In computer science, I would not teach anyone that a variable possesses at a given moment, one value and one. The processing of natural language, undertaken for the purposes of machine translation, has for a long time stumbled on the intrinsic difficulties of artificial intelligence, and particularly on the modeling of common sense.

What can be said about the distribution of these points in the plan? It also reads, with regard to the training of future computer scientists: A general education in mathematics and experimental sciences is essential so that they can permanently find their place in a rapidly changing world. Because it is very rare to find in one person the mastery of these sectors of trades. On the other hand, to provide an example, we may also use mathematical and computer techniques and provide a work program in the course of organization.

We recommend the study of a program in this direction. On the other hand, to provide an example, we may also use computer science, mathematics, and mathematical logic for the establishment of databases, to provide an example: types of clientele, identification of clients at risk, etc. The operating procedure for the coding setting up the activity is important.

By expressing that the sum of the squares of distances which one will specify is minimum to determine the expressions of a and of b then their values. A paradigmatic example is the recipe for food, which solves a problem: make a food item. No one can predict in detail the changes, but the movement of living together is clear and its repercussions in secondary education are certain.

But a traditional strong point of computer science, mathematics, and mathematical logic teaching is the importance given to abstraction. We may also provide probabilistic techniques optimization, Hadoop, MapReduce, and analysis of aerial or satellite photos. On the other hand, to provide an example, we may also use computer science, mathematics, and mathematical logic for statistical consulting companies. To illustrate, we may additionally supply and help with the strategic decisions of companies or the commercial marketing decisions.

Support of rhetoric, it remained a long time a separate discipline of mathematics, before the considerations on the foundations and mathematical notation developed a renewed discipline called mathematical logic, including the theory of demonstration. The programs, in a justified way, highlight the coherence of the mathematical notions, and this coherence, very strong, tends to conceal the place of the mathematics within the other sciences and the culture. On the other hand, to provide an example, we may also use mathematical and computer techniques for mathematical notions.

There are thousands of mathematicians, it is a huge science and no other can compete, let alone the number of awards or the number of institutions involved in mathematics, formal calculation, and constructive mathematics. Conversely, mathematicians use computer visualization in geometry or dynamic systems. They brew concepts some of which come from the bottom ages, such as numbers and figures, others appeared few centuries ago, as functions and ways of representing them, groups, probabilities, differential equations, and others still appear, like everything that comes from the computer.

The objectives are as follows: discover the notions of correcting code, make the need for a coding convention emerge, work on the concepts of coding information, exploit the basics of numeration, discover the notions of compressor code, etc. Mathematicians made it possible to express in a rigorous and transmissible way concrete phenomena. On the other hand, to provide an example, we may also use the work on the influence of computers and computers on mathematics and their teaching. The results of the simulations make it possible to adapt the development methods of products placed on the market in order to increase their safety and reliability of operation.

It seems to me that we can only agree, in general, with these general considerations on the links between computer science and mathematics, although, on the contrary, we are interested only in mathematics. Already, mathematical work which is not limited to calculation in connection with computer science can also be seen as a way to increase interest in mathematics for students in which it is weak. There is no job that goes without mathematics!

On the other hand, when it comes to combining skills in economics (macroeconomics and pure and political economy) and statistics, it is most often necessary to create mixed teams of macro-economists and data analysis specialists. Now it is ultimately thanks to these achievements that the general public can now have the web, DVDs, Internet, mobile phones, etc. Their implementation shows the importance of thinking deeply about their articulation in mathematics education, mathematics, and mathematicians. Originally, mathematics was the science of figures of space, numbers, and movement. The Second World War, to provide an example, saw a period during which the mathematician John von Neumann plays a role known to all.

They develop their tools, their methods, their concepts from their fields of study, to which they refer to test the validity of their results. On the other hand, to provide an example, we may also use mathematical and computer techniques and provide contact and composition of the groups. The group will begin by appropriating the work already carried out and explore the interactions between computer science and mathematics and their interest in teaching.

It is, therefore, a real programming tool on paper. To illustrate, we may additionally supply a strict order that is an antisymmetric, irreflexive, and transitive relation. This shifting regulation based on international and US standards encourages the development of elaborate home models for risk assessment. On the other hand, to provide an example, we may also use mathematical and computer techniques in marketing services and provide a study of consumer behavior in order to establish typologies of customers to adapt the service products sold.

We refer to the forthcoming reports on higher education for the examination of the situation of the first cycles. In the face of the computer, do mathematics change in nature, or do they change more punctually only where new knowledge appears? To translate it into the whole of education, it imposes new tasks on professors and researchers, which must be recognized by the universities.

At the end of these years of toil, efforts are paying off. The applications in this field are intended, to provide an example, to: provide large consumers of numerical calculations; answers can be provided in the design of vehicles to improve the safety of motorists; simulate crash shocks and the resulting body damage by using mathematical modeling incorporating biological, mechanical, physical, and medical laws; predict the resistance of structures, materials and their deformation: to provide an example, the tire or the ski by resorting to mechanical modeling; etc.

Computer innovations, such as those of logic and mathematics, appear at the whim of a story that often stumbles or stutters, but they do not conceptually depend on their circumstances of appearance. Let's first say that, in the face of Moore's Law, the doubling of accessible or storable information currently obeys a growth rate of a factor of two every 9 months or so. Also, we live in a period where a student can acquire a data storage device to maintain the equivalent of several tens of thousands of books, and when a low-cost laptop has a computing power a million times greater than that of the first electronic calculators.

Another way of approaching this problem is to extend the verification of programs, traditionally carried out by a battery of tests, into a symbolic verification in a more abstract computing space-called model checking. We will justify these answers. To provide an example, bio-statisticians are highly sought after in the labor market, whereas adapted training courses produce only a very limited number of them at present.

The candidates for these mathematical and computer techniques should concretize with students the thematic of the information theory, that is to say to propose a reflection around the notions of data exchange, information quality, encoding of information, treatment of information, algorithmic quality, etc. On the other hand, to provide an example, we may also use mathematical and computer techniques particularly for the software industry, and highlight the decline of and, in particular, of the (despite some notable exceptions cited in the text) for years, in these sectors.

On the other hand, to provide an example, we may also use mathematical and computer techniques master the modular arithmetic tools useful for algorithmic quality. Machines include computers, more specialized devices such as cameras or phones, and more generally all physical systems, for which we have defined an interaction protocol, which allows us to exchange data. Students are therefore baffled. To illustrate, we may additionally supply combinatorial analysis: combinatorial counts and transformations associated with the series algebra.

This is the image of university education at the undergraduate and postgraduate level when it is addressed to future mathematics professors. In the needs analysis, the focus is on the strategic character for a country's computer research economy and related industry. Probabilities are an integral part of mathematics, we see it every day. In the mathematical and computer techniques, research in language theory was extended to systems for describing program schemas and abstract syntax trees, as well as functional notation systems and calculus.

Of course professors can teach the necessary mathematics. On the other hand, to teach the necessary mathematics the duration varies in general. Therefore, a problem of algorithmics is very specific. To return to the place of these sciences, indeed, the statistics and the probabilities, on the one hand, computing, on the other hand, have developed enormously.

On the other hand, to provide an example, we may also use mathematical and computer techniques for mathematical tools applied to finance, statistics, stochastic processes, probabilities, etc. To illustrate, we may additionally supply courses of geography are mines of work possible in mathematics: populations, lines of level and potential, basin of attraction of the rivers and fractality, statistics, etc. We recommend that, in particular, at the level of the college the programs plan explicitly from and, on the other hand, the articulation of geography and mathematics courses. To illustrate, we may additionally supply and link computing to the real world and other disciplines. On the other hand, a convenient way to teach a little computer science is to limit yourself to simple mathematical examples.

To provide an example, you may wonder "Does this code work for any example?" When the level of education at a discount I would say rather that it is a bit too high, to situate me precisely in those who were recalibrated and who had the urge to go explore more interesting topics after, which is yet considered as something easy by some. On the other hand, to provide an example, we may also use mathematical and computer techniques as a Database Administrator. In a company, the databases often constitute a concentrate of information capital, even strategic.

Thus, the first applications are largely the result of digital analysis of differential systems. A dual skill is needed here in management and computer statistics. To provide an example, we may also give an overview of the history of computer-assisted mathematics, a story shaped by the proximity and distance between the user and the computer.

Since the early days of computer science, the historiography of computer science has gradually diversified and enriched, in connection with the gradual diversification of uses of the computer. On the other hand, to provide an example, we may also use combinatorial optimization, which is largely the result of computer processing of operational research problems, links to complexity theory through the works. On the other hand, computer use of the computer tool through the power of calculation and statistics-probability coupling in random universe, is also useful.

It is a question of characterizing the behavior in number of elementary operations, of these methods of computation. Thus, we can imagine a machine in which we drop a ball in a vacuum for a time and measure the distance traveled by the ball. These suggestions are intended primarily for first semester students in Mathematics or Computer Science. To illustrate, we may additionally supply and synthesize a correct certified program based on the proof of coherence of his specifications.

Does not his discipline carry the name of computer science? These artificial intelligence systems are therefore theoretically viable; many people even had fun experimenting with it. Artificial intelligence is probably even more mathematized than anything else. The articulation of the mathematics and physics programs has been the subject of good work, in particular, to link exponential function. Admittedly, meetings and roundtables have been very numerous, but without clear conclusions.

The connection between elements is known gradually. It is thus possible to predict a considerable development of the applications of natural language processing in the next decade. It's the same thing here, you think that computer science does not require mathematical knowledge because you're doing it while being bad at math.

On the other hand, the so-called analytic approaches push computation to its limits in the discrete domain on the basis of combinatorial analysis, then remove the information sought at the cost of an interpretation of the generating series in the complex domain and a fine calculation of the associated singularities. The Second World War, to provide an example, saw a significant investment in the development of computers for military purposes anti-aircraft defense but also cryptographic analysis. The volume in steady growth of the diversity, the data to be treated, and the heterogeneity of these same data require methods of rapid access to non-digital information.

On the other hand, to provide an example, we may also use scientific computing, bidisciplinary skills, simulations, mathematics, and computer science. The skills required here cover mathematics applied to computer science, insurance probabilities, finance probabilities, insurance statistics, actuarial statistics, actuarial probabilities, finance statistics, mastery of database software, etc.

On the other hand, to provide an example, we may also use mathematical and computer techniques for computing, medicine, mathematics, statistics, etc. Strong skills in mathematics and computer science provide a background of fairly extensive mathematical culture. To provide an example, If a and b are divisible by c, then a + b and a – b are divisible by c.

On the other hand, this research makes it possible to design sophisticated programming environments, assisting a designer to build software that conforms to his specifications. I'm just sorry to see that with politicians willingness to put maths everywhere and for all politicians have killed public opinion towards mathematics. There are seeds of innovation at all levels. Mathematical and computer techniques for medicine, to provide an example, is a market that brings together several disciplines (mathematics-computer science and medicine), on which we lag behind.

On the other hand, to provide an example, we may also use data mining, data analysis, statistical skills, etc., and computer science for mastering database management. To illustrate, we may additionally supply and indicate a reduced equation by specifying the associated reference. I think we've gone too far, but I think we'll finally come to a balance. Remember the algorithmic, which systematizes the notion of calculation, and logic, which formalizes the notion of demonstration.

Discrete mathematics are by definition the mathematics associated with such finite or countable structures, and are therefore a priori the closest to computing. Similarly, the growing demand for mathematical tools in biological models has led to the creation of mathematics, biology, computer science, etc., programs. On the other hand, to provide an example, we may also use numeration and conversion and provide several ways to convert an integer to the base 2 (binary) or to the base 16 (hexadecimal).

It is on this ground that will develop what is, first of all, a set of technical know-how and which will then give birth to the computer science. If we start computer science, this concept is well before computer science, but if we start mathematics in the fourth century before our era, it is also prior to mathematics. In the positive case we will indicate a diagonal matrix similar to as well as the corresponding base change matrix, after that depends on your training.

As a result, a number of examples can be at the border. This need is, of course, no problem when working on paper. This is sometimes called randomization and is introduced voluntarily in the calculation. The proposed conceptual level often exceeds that usually found in mathematics and the necessary time would be considerable. On the other hand, to provide an example, we may also use mathematical and computer techniques for diverse fields and large companies belonging to various sectors.

However to answer you directly it is true that everything is not to take but overall the math still imposes a certain methodology and a good approach to solving problems. In addition to the programs and competitions, a review of the mathematical and computer techniques, which is a possible germ of interdisciplinarity in the preparatory classes, should be made. The event that constituted our realization that these concepts formed pieces of a puzzle that fit together perfectly, and the transformation of the science that it involved, was an event a priori very unlikely, an event as there is little in history.

Numbers, figures and functions describing movements are studied respectively by geometry, algebra, and analysis, which are the three main branches of mathematics. Computer science has roots that go back to ancient mathematics, through main streams. We therefore immediately understand this mini-language.

Another avatar of denotational semantic studies, initiated by theoretical computer scientists to understand the notion of possibly non-terminating computation, was the development of real-time programming, notably around reactive languages at the laboratory. And that takes us to another point, the discount requirements in the education level, today everyone can be an engineer, even having average scores in math, but that's another debate.

On the other hand, to provide an example, we may also use mathematical and computer techniques for all levels and on all subjects (architecture, economy, etc.). To illustrate, we may additionally supply an algorithm we implemented as a rhyme so that his brain keeps it without worries.

We may also use mathematical and computer techniques for modelization, statistical forecasting method, actuarial statistics, etc. Students with mathematical and computer techniques qualifications are highly sought after in many fields such as: all areas related to the notions of imaging, cryptology, computer networks, graphical computing, social networks, robotics, computer security design, analysis of algorithms, telecommunications, networks, etc. Roughly, a fine and original analysis of the combinatorics that is at play reduces the infinity of the possible exceptions to a high but finite number of special configurations.

On the other hand, to provide an example, we may also use mathematical and computer techniques for logistics and production management. To illustrate, we may additionally supply what tariff policy must be applied to ensure the optimal filling of airplanes? Unforeseen and chance have an important weight here.

The theory of reactive languages, which was already authoritative for the design of control automata, finds a new field of application in the design and verification of circuits. On the other hand, to provide an example, we may also use mathematical and computer techniques for meteorology. The weather forecast remains a major scientific challenge. The phases of assessment are therefore important, as it is at this time that students will confront the different codings.

The determination of the optimal trajectory of a probe, the launch and the maintenance in orbit of a satellite require a large consumption of calculations and mathematics. I find it unfortunate that they were so important in my day, because I am willing to bet that it has unnecessarily closed the door to computer science for many students. Excluding programs, the supervised personal works were the most visible display of the will to develop interdisciplinarity.

Engineers, chemists, physicists, etc., have built, developed and used numerical and analog computing systems for both practical and mathematical purposes. They also demonstrated important theorems using large computational calculations: Kepler conjecture theorem on the arrangement of spheres, four-color theorem, etc.

On the other hand, to provide an example, we may also use mathematical and computer techniques for the transmission of information. From new technologies to health, research, banking, teaching, finance, industry, consulting, services, etc., these issues affect professionals and different sectors. The same goes for maths, it's an elementary baggage. As it is a sequence of actions, his brain, like all human brains, can easily learn it "like a dance or a song."

In one theory, there is as much true science as there is mathematics. And we manage to demonstrate it. On this point it seems to me that things are progressing reasonably but still without really a debate on the programming part.

In addition, precautions such as a priority system allowing an encryption of the message if it contains sensitive data should be surrounded. They have, even before learned societies, taken initiatives along the lines of this report, by broadening their contacts and their actions beyond mathematics.

Mathematics and computer science provides a background of rather broad mathematical culture. Computing will never give anyone any understanding of mathematical concepts. On the other hand, to provide an example, we may also use mathematical and computer techniques for operational research techniques.

To provide an example, if the b's are frequent in a series, but the a's and the d's are rare, we can decide to recode b afterwards 0 and a, c and d by the sequences 10, 110 and 111 we obtain then a shorter sequence than the initial sequence, from which we can, however, reconstruct it. Moreover, their movement is as fast as that of other sciences. Computer skills therefore correspond to the mastery of heavy statistical software.

On the other hand, to provide an example, we may also wonder 'Why go to the doctorate?' NP belongs to this category of questions. To illustrate, we may additionally supply and allow a massive exchange of extremely varied and unstructured data.

On the other hand, to provide an example, we may also use mathematical and computer techniques for applications that require: solving a complex system of equations and finding equations that describe the evolution of atmospheric conditions.

Natural language formatting is part of the mathematical activity. Apart from its unusual hardware support, a notebook or a blackboard, it has all the features of a computer language. To illustrate, we may additionally supply and develop by a working group whose constitution was announced. An accurate evaluation of the device would be necessary to determine what are the conditions necessary for a successful implementation accompanied by a major effort of initial and continuous training.

On the other hand, to provide an example, we may also use mathematical and computer techniques for a study of finite automata and graph algorithms Associated is one of the national fields of excellence, with laboratories and a computer lab. So we work only with a digital reflection of the real system with which the algorithm interacts.

Scientists know that these algorithms have become structuring tools including in the sciences of earth, life, matter, etc., balanced between mathematics and computer science that allows us to train students attracted by the theoretical computing, mathematical aspects, or new applications of algebraic and geometric disciplines and discrete mathematics. To illustrate, we may additionally supply and specify in all cases if the polynomial admits different roots and for a positive answer we will give an example of root.

On the other hand, to provide an example, we may also use mathematical and computer techniques for job opportunities: signal processing, network engineering, cryptographer, statistician in a bank, it management of companies, etc. To illustrate, we may additionally supply techniques that were just written, so there will inevitably be errors. Such a use of tools seems natural, since an algorithm is made to be executed without thinking and this is how, since antiquity, we used abacuses, charts, chessboards, etc.

To illustrate, we may additionally supply and show that an abstract framework is extremely effective in other disciplines: it is to implement a little modeling. On the other hand, to provide an example, we may also use mathematical and computer techniques and recommend to the doctoral candidate to take advanced courses and to participate in seminars. If the applications can serve as motivation or illustration of a well-constructed course, they should not eat too much space compared to the theory.

Remember, every cube of integer is a sum of no more than nine cubes. Conversely, one learns a formal language by first learning his grammar. In class, it is not enough to check it by measuring, as one day a professor suggested to me, it is absolutely necessary to prove it. Examples abound in with the increase of information flows and the need to decide in real time, computer scientists have had to offer new tools such as integrated management software or customer relationship management.

To illustrate, one of the big differences between a nurse's aide and a nurse is precisely in the calculations of flow rates, treatment part, doses, etc. Notions that you have already seen but which one will need (and that it is necessary anyway to control when one does mathematics or informatics). In terms of the objects it studies, computing is approaching mathematics, because both are interested in abstract objects.

Professors in other subjects are hardly prepared to turn to mathematics professors. It will be necessary that our border community every day a little more permeable (the situation is stimulating), invest more. Help in the selection of new services. To illustrate, we may additionally supply help for the subjects of languages, geography, history, philosophy, etc.

It will also be of interest to students in computer science, or those in preparatory classes wishing to acquire the basics of algorithmics, as well as all those who wish to know and master the mathematical tools necessary for a good practice of computer programming. Research is now very active in mathematics. The object of study of computer science has gradually moved from machines to algorithms and languages.

I would be tempted to classify mathematics at the same level as the Design Patterns: I'm sure that for the most part, the comparison may seem flawed, but imagine trying to teach Design Patterns in a purely theoretical way to pass the tray. International surveys show that mathematics, computer science, and mathematical logic students are uncomfortable with modeling problems. It takes years to operate in these disciplines.

The computer discipline, whose presence is essential to this reflection, should therefore be represented by the society and the associations public education and computer science. For example, the term computer variable is actually bad choice. On the other hand, writing an integer in base 2, 10 or 16 is unique.

Perspectives in preparatory classes allow for the implementation of interdisciplinarity at the same time as student initiative. I came to wonder if we were asking the right questions, hence my introduction of the rest. But the interaction took a long time to get into place properly; indeed, set theory and, in particular, the efforts would stiffen the official discourse of mathematics around a standardized first-order set theory language.

The microprocessor of a computer only works with two digits: 0 (no current) and 1 (current). To illustrate, we may additionally supply a program written in a high-level language designed for humans. On the other hand, we may add mathematics and computer science in the collection, to provide an example, and use petri nets.

41 CHAPTER

PROGRAMMING PERSPECTIVES

The idea of the sector was to make interactions between different subjects. This allows remote communication and data exchange. A famous conjecture of the nineteenth century states that any map of geography (that is to say, any planar graph) is colorable with at most four colors. However, the companies cited point out that there are few places where high-level application-oriented statistics are taught.

As to whether this is the role of the math professor, that's another question. That's why a lot of students leave to work right after the license. And then the professor must not live this as a constraint.

On the other hand, to provide an example, we may also use formal tools and mathematics applied to computer science. But it is often the requirements of other disciplines that lead to forging an adequate mathematical tool.

To provide an example, we may also remember that private and public laboratories invest a lot in research to understand the mechanisms of life, agricultural issues (e.g., plant breeding), and to respond to medical issues (the treatment of genetic diseases).

On the other hand, to provide an example, we may also use computer science, mathematics, and mathematical logic to provide patent evaluations, and a possibility of replacing with individualized help. To illustrate, we may additionally supply and introduce computer science, mathematics, and mathematical logic without specifically targeted means and led in the vast majority of colleges to an hourly decline in several disciplines, often the most.

(and reluctantly) acknowledged, cryptography is a need of the general public.

On the other hand, to provide an example, we may also use computer science, mathematics, and mathematical logic and a number which is not prime is divisible by a prime number, and can even be written only with prime numbers, in the form of a product. Flows are significant and particularly important for manufacturing processes in order to respond to issues such as just-in-time management, quality control, control of the hazards of production lines. Image processing?

We can then question them in this way: "Are there better codings than others?" I always prepare very seriously the computer science, mathematics, and mathematical logic courses, much more than the mathematics.

As we have pointed out in a few selected examples, the mathematical approach at the same time imbues the development of fundamental computing: a number of theories and mathematical results are used, while the mathematical schema definition-theorem-proof plays a role and capital in structuring knowledge. The same goes for my case: mathematics was not my forte during my training, and I am come out today in the computer. Would the result be brighter than the results in math?

42 CHAPTER

ADVANCES IN LINEAR LOGIC

These disciplines are sought in a complementary way in applications concerning the research and development sectors of firms. I think we need to quickly discover the ways of learning that best correspond to us.

As soon as we want to make transformations (whether images, objects, or anything) we will use matrix calculus, changes of benchmarks. On the other hand, to provide an example, we may also show that at that time an important part of the teaching services of university mathematics professors was at the service of other disciplines, and that it was the source important innovations. To illustrate, we may additionally supply disciplines that have become more and more intelligent thanks to embedded systems.

On the other hand, to provide an example, we may also use computer science, mathematics, and mathematical logic to systematize the algebraic investigation of a hierarchy of formal languages, whose regular (or rational) languages form the first stage, and Chomsky's non-contextual languages (or algebraic languages) form the second stage.

On the other hand, to provide an example, we may also use computer science, mathematics, and mathematical logic as an expert in Internet and multimedia technology, image processing technician or computer maintenance, telecommunication engineer, researcher finance, banking, insurance the it management deals with accounting, r & d engineer, billing, inventory management, optician, acoustics engineer, etc. There are several courses of study. The system could even be generalized, with some inconvenience: it is possible to use a large number of data without significant interferences.

We often talk about "digitization" to designate this operation, because by noting 0 for a black square and 1 for a white square, we can write this continuation 11001011. This gives an approximation of the image: to provide an example, tiny details smaller than the size of a square are lost. What we already find, and which remains to a large extent a mystery is, according to physicists the unreasonable efficiency of mathematics in this sector.

to provide an example, universities are offering biologists the opportunity to complete their training in computer science to specialize in bioinformatics. It is out of the question to advocate a single model, but it is interesting to observe the systems of other countries and to take them into account.

Another advantage would be to show students in computer projects that without mathematical concepts (admittedly discovered and implemented in a different setting), we are very lacking! But to look at everything that can be delegated to an algorithmic mechanism to, beyond, allow and force the human mind to surpass itself. On the other hand, to provide an example, we may also use systems that can simulate the growth processes of a tree, in successive stages, according to time scales and spaces.

I think it's good and healthy to do calculations. But the important thing is not really the question of contents. However, this type of machine can solve only a finite number of problems: even a knife of the Swiss army has only a finite number of blades.

These disciplines are often synergistic in the medical and biomedical fields. The answer to the preceding questions should make the subject much less subject to any debate, finally it seems to me and that following is my personal opinion, and, it is the one to which I am most inclined to rank. We have cited some excellent examples and some difficulties.

But it has nothing to do with it, it is only a question of an individual blocking: Maths is only valid for you if they are applied concretely, do not believe that you do not do maths in computer science. Strong skills in mathematics and computer science provide a broad mathematical background. As the class of groups of finite rank is strictly greater than that of algebraic groups, we will also see the new phenomena which appear in the model-theoretical framework in relation to it, the more restricted, the geographical anomaly.

On the other hand, to provide an example, we may also use computer science, mathematics, and mathematical logic to query volumes of data organized according to different criteria. This discovery, made at the turn of an exercise, is not obvious: it took centuries for humans to release this abstract notion of function. The history of algorithms is nonetheless a point of encounter and dialogue between the history of computer science and the history of mathematics. The link between mathematics and other disciplines must be reflected in the fact that all courses for mathematics students have a historical component and a serious introduction to a different scientific field.

When statistics are taught in mathematics, it is good to see different aspects in geography, economics, physics, etc. The introduction to statistics lends itself to joint training of professors from different disciplines. On the other hand, to provide an example, we may also use and encode information that we understood.

You may have done math for years, at some point you will not have the resources to meet the needs of the trade, and the books, or a true business expert will become your best friends. I only knew the functions defined by formulas, and I wondered how I could deal with this problem without formula. On the other hand, to provide an example, we may also use probability-statistics applied to the theory of information.

It is difficult and it requires a regular effort of continuous training. It would be to forget his roots. And if it's still the case, it's very unfortunate. The mathematical tools here make it possible to predict the temperature levels reached by these parts and to consider when, and in what way, it is necessary to intervene to cool them.

It can also design protected communication systems, in the field of digital transmission security. The results will be justified. On the other hand, to provide an example, we may also use a computer simulation of accidental shocks and the resulting damage to the body, to provide answers to improve the safety of drivers and passengers.

They will be an opportunity to consolidate and deepen students' knowledge of the use of measurement units and the practice of certain unit changes. But if the history of computing has undeniably inherited history of mathematics, we can also today ask the opposite question: can the history of computer science inspire the history of mathematics? On the other hand, to provide an example, we may also use statistical tools and probabilities.

Let's take an example to conclude this section. To illustrate, we may additionally supply computer science, mathematics, and mathematical logic and send digital photos by mobile phone. The characterization of decidable subclasses, and the complexity of their membership, their equivalence, or other algorithmic problems, is a classic theme since the beginnings of theoretical computing. Certainly, over time the computer has emancipated from its humble origins to become a machine with multiple faces that performs a function by its location in an information network and by the footprint it makes on the flow of information that pass.

The consideration given to professors is an important issue. On the other hand, to provide an example, we may also use computer science, mathematics, and mathematical logic for the independence of the network layer from the physical layer. On the contrary, it is not adapted to a shift register that does not record the figures obtained: it is precisely a matter of not burdening the memory of the computer with random numbers but of generating them ad hoc when we need it. It also allows hesitant students to delay the choice between these two disciplines. To illustrate, we may additionally supply computer science, mathematics, and mathematical logic to animate reflection among mathematics professors, and to create interdisciplinary links.

On the other hand, to provide an example, we may also use computer science, mathematics, and mathematical logic to provide a matrix that can be considered as the matrix associated with a linear transformation L defined in a vector space related to the base B. Promising interfaces emerge: ecology, topology, databases, genomics, structural biology, etc.

To illustrate, we may additionally supply an advantage is that it is algorithmic, so programmable. Their processing consists in analyzing their structures to compress them, to decode them or extract the noise to confront them with dictionaries of signals or test images in order to classify them or to look for the keys. On the other hand, to provide an example, we may also use computer science, mathematics, and mathematical logic for the growth of functions, sequences and summations, sets and functions, sets and operations on sets, predicates and quantifiers, logic, equivalence, etc.

On the other hand, the reduction of polynomial systems is related to contemporary algebraic geometry, the symbolic manipulation of many special functions of physics becomes possible thanks to the joint use of modern differential algebra and adapted data representations. Polynomial factorization is best done by methods that exploit both the structure of the finite fields and the principles of randomization. On the contrary, it is good to start from all that is familiar to access the simple and the general. On the other hand, to provide an example, we may also use mathematics, (mainly numerical computation) applied to the problems.

Let's try to say why and how. At the same time, expectations are becoming more complex for users who want open systems integrating software, products, and services. However, it is not the fact that the information is expressed as an integer that is important, it is the fact that it is expressed symbolically.

On the other hand, to provide an example, computer scientists should also meet a few times in the year, and computer scientists may invite experts, and set deadlines to produce a first computer science, mathematics, and mathematical logic report. So they have a little trouble understanding that their students will not necessarily follow this path too, I have the impression.

to provide an example, we may use computer science, mathematics, and mathematical logic and also provide a reorientation and bridges between sectors. To provide an example, security products may include: smart cards, copyright protection, mobile phones, wifi, biometric identification, digital image, document marking, etc.

The search for information in large corpora, such as the Web, and the organization of corporate memories, are major economic issues using these techniques. First, it must be a constructed course, where we start from basic postulates and demonstrate the tools that will be useful. First, computer scientists explains very simply an algorithm is simply a way to describe in great detail how to do something.

So here we are creating a minimal language to code this information which we understand him and me the meaning. Other similar formalisms, like Milner's calculus for communicating systems (CCS) as a process, were later invented to model the competing processes. The academic mathematicians mentioned above reacted in the same direction, the final draft should therefore be significantly evolved, without being able to fully satisfy us.

Finally, we propose to consider computer science as a formal science that one can think without a computer. Their success is confirmed. On the other hand, to provide an example, we may also use computer science, mathematics, and mathematical logic for decision-making purposes, scientific data from the laboratory, to obtain authorization to market a new drug, clinical trial results, etc.

Through his relationship with mathematics, we have illustrated the fact that computer science is a science very much in the tradition of other sciences. The corresponding work can only be conceived from concrete situations and in conjunction with other parts of the program.

Classes are linked together by lines representing their associations, that is, the relationships that allow them to interact. Research institutes therefore provide some formal demonstrations that are only algorithms to apply, and in this math approaches the computer.

On the other hand, to provide an example, we may also use interdisciplinary activities where it is necessary to examine separately the programs and the reality of teaching. One day, maybe that same person will end up in a company doing R & D and being asked to write a math-related program.

This is stated by the following property. It appears that some vocational colleges have found a framework for real interdisciplinary activities where mathematics have a recognized place. The teaching of mathematics in relation to other disciplines. It turns out, to provide an example, that the best big integer multiplication algorithms involve a Fourier transformation in a discrete version.

Here, we must specify the version of the protocol and other information, the most important of which are the source address and the target address. I think that basically computing is logic. On the other hand, to provide an example, we may also use an institutional place of interdisciplinarity at the college.

At the end of the 1st session, the activity is linked to the notion of binary coding of information. I am therefore a perfect example of the fact that it is perfectly possible to be a mathematician and yet perfectly able to succeed in a professional career in computer science.

On the other hand, to provide an example, we may also use computer science, mathematics, and mathematical logic for a dictionary algorithm. Searching for a word in the dictionary seems to be the easiest thing in the world; but who knows how to do it effectively? to provide an example, we may also use computer science, mathematics, and mathematical logic to provide characteristics of recruitment in different sectors of opportunities, to provide an example for the consulting and engineering sectors. This sector includes services companies, service providers in the field of imaging, pollsters, remote transmission of information, consultancy firms, etc. After all, if you limit yourself to creating websites, the problem is much less obvious.

We will show here, conversely, that a non-negligible part of the concepts and methods of the computer discipline is independent of the world of computers. And, once their requirements have been modeled in UML, they can transmit their schemes to computer scientists.

I was a student in the United States and I could compare both systems. On the other hand, to provide an example, we may also use computer science, mathematics, and mathematical logic for a positive case, and indicate using as few calculations as possible, a diagonal matrix and the corresponding base change matrix. What about mathematics?

All computer science comes from mathematics, and we only notice it when we start contributing to very advanced areas of research. On the other hand, to provide an example, we may also use computer science, mathematics, and mathematical logic and apply various methods of graphs and trees (including infix, prefix, postfix and tree paths)

The modulation, according to the grade level, of these objectives is supported by the analysis of the specific character of the computer science, to distinguish from the learning of the use of the numerical tools, the one and the other to be present in education. It is a recommendation, already made in previous reports that we resume here. We will justify these answers.

Just as certain precautions must be taken in set theory to avoid the paradoxes resulting from too lax set-lists such as {x x x}, it is advisable to provide the calculus of stratifying limitations to restrict it to convergent calculations. To understand, take the example of and the base 10. More generally, we have the following property. Even if Google gives me some elements, it will never be enough unless I have, of course, the next years and the necessary patience that goes with it, all that to re-invent things that mathematics have already solved since a lot of time.

The computer programs thus become skeletons of the proof of their adequacy to formal specifications. The formal calculus integrates the data structures and general algorithmic of computer science in an original way, and various "divide-for-rule" recursive sharing methods (the so-called "fast" FFT calculation, is the Cooley–Tukey algorithm), and the adapted use of arithmetic congruence properties.

Until the nineteenth century, great mathematicians, such as Gauss, Leibniz, Newton, or Euler, invented original methods of numerical and symbolic calculation. How to become an engineer? Circuit design research has itself generated a wide variety of formal methods – the size of integrated circuits no longer allows the use of traditional methods, and circuit verification is now integrated into the design phase.

When we write $y = 1 + 2x$ the "variables" x and y take an infinite number of values. The history of mathematics education in colleges shows that its relations with the teaching of other disciplines have varied over time according to the types of training and the tensions between theoretical and practical training. And this not only because this algorithm was a good match to an existing technology, but also because it was possible to develop the mathematical theory of the Fibonacci Sequence.

And to be able to learn other things, you have to use the abstract. The bad, on the other hand, was the discovery of the functioning of the administration of the faculty. The effort to implement these recommendations, and if possible to go beyond them, involves all national education and the entire mathematical community, and requires a variety of competitions.

The scientific collaborators of the sections of Mathematics pursue researches in very varied fields: von Neumann algebras, functional analysis, dynamic systems, statistics and probabilities, numerical analysis of the differential equations, algebraic topology, geometry of the groups and combinatorics, algebraic geometry, combinatorics, node theory and singularities, arithmetic and algebra, etc. Moreover, it is not possible to really get into the operation of a computer, it would be necessary to teach a little assembler. On the other hand, to provide an example, we may also use computer science, mathematics, and mathematical logic for competencies quite multidisciplinary.

More than ever they feed on other sciences, and they feed them. This led us to renew the analyzes of the mathematical structure of chaotic attractors, by coupling the calculated images to the geometric and topological approach. The same goes for a result in computer science.

Like them, they can be valued by direct applications (this is the case, in particular, of statistics). professors teach things they have never learned in their initial training.

Although the technical things seem to be randomly ordered, the sequence obtained becomes in reality periodic after terms. It is on this ground that will develop what is, first of all, a set of technical know-how and which will then give birth to computer science. This thought has, of course, many precursors in history, such as Galileo who introduced both the use of instruments, the telescope, and proposed to use a new language, the mathematical language, in the natural sciences but what seems new, however, is to pay attention simultaneously, in the same gesture, to these questions.

On the other hand, to provide an example, we may also use computer science, mathematics, and mathematical logic for modelization and methods of decision support by intelligent data processing (example: "Data mining"). I agree that we all have our limits especially with a college. To illustrate, we may additionally supply and determine the local extremums of these functions and their nature.

43 CHAPTER

SYMBOLIC LOGIC

On these questions, a set of methods emerged: we came up with the concepts of data structures, then databases, and we simultaneously identified some major paradigms of programming, such as the famous principle of divide-and-conquer algorithms based on recursion. With this theory, we also give a practical approach to the notion of "random nature" or "randomness." Before the activity, few students know how a computer works and this activity introduces the binary coding of information in a devious way.

Especially that to make the web maths will not serve much. These languages are programming languages so simple that one can conceive of electronic machines capable of executing programs directly and to which one manages to translate and to compile and often in several passes, the programs written in the languages of higher level, that is to say more ergonomic. Modeling and simulation therefore apply to industrial processes to describe the predictive operation of these processes that will be induced by the modification of sensitive parameters.

There are almost no examples either: there is the lecture for that. On the other hand, to provide an example, we may also use mental math questions on the way to college. To illustrate, we may additionally supply computer science, mathematics, and mathematical logic and could be done on the occasion of the release of a report on mathematics in the contemporary world.

On the other hand, to provide an example, we may also use modeling to help with the decision, (e.g., evaluation of the elimination of certain calculations). We are very far from it, and the gap widens between the science not for us to judge the feasibility of the proposals made for this curriculum, often very detailed (especially for the primary), nor their magnitude.

Still, the programs proposed by the computer scientists are at the same time very ambitious, diverse and often imprecise. The Web changes the game! It is legitimate to enroll them in college programs, but only if the objectives I have stated are respected.

They have only a few choices. A more algebraic characterization, in terms of rational languages, represents the languages recognized by these automata as the sets of words generated by operations of union, concatenation, and iteration, the regular expressions. On the other hand, to provide an example, we may also use the field of speech processing that is in the process of providing human-machine interfaces using voice, usable by speakers without learning in restricted semantic domains.

It was a discovery for the computer science, mathematics, and mathematical logic and a speech quite commonplace. On the other hand, to provide an example, we may also use software tools to provide to and to be an interface with measurement software for diagnostic assistance, an image acquisition system (MRI, scanner, etc.), 3D organs visualization software, etc. The great ignorance of students and their professors about these summits of thought accentuates the dehumanized image of mathematics, but has far more serious consequences for the future: that of allowing the gap between advances in science and the common culture of citizens.

In all the missions entrusted by the banking-insurance sectors, these various competencies are used in a combined way, with a different weight for each according to the mission entrusted to them. On the other hand, to provide an example, we may also use computer science, mathematics, and mathematical logic for the tire and a test of resistance of the materials to the deformation. To provide an example, we may also use computer science, mathematics, and mathematical logic to provide an arithmetic base 2, also known as a binary base, that can be counted (only 0 and 1 are used): 0, 1, 10, 11, 100, 101, 110, 111, 1000, 1001, 1010, 1011, 1100, 1101, 1110 and 1111 and so on.

On the other hand, to provide an example, we may also use computer science, mathematics, and mathematical logic for products, to describe what will be the projected characteristics of the final product made from a variable combination of components. Computer science also plays a role in the life sciences, by making it possible to process with a very often very fine algorithm the masses of data that are presented there and to extract the essential information, it is the domain of bioinformatics. To illustrate, we may additionally supply and model various real-world problems encountered in computer science using the appropriate forms of graphs and trees, to provide an example the hierarchical organization of files, the representation of the topology of the networks, etc.

However, on the whole, the contrast is surprising between the very strong interaction of mathematics as science with other sciences, and the little relationship of their teaching with other scientific courses. We will first identify the main historical roots rooted in the classical scientific tradition.

I'll let you imagine what difficulties I may encounter. On the other hand, to provide an example, we may also use computer science, mathematics, and mathematical logic for information processing surveys, statistical institutions, advertising resources, engineering consulting, etc. To illustrate, we may additionally supply computer science, mathematics, and mathematical logic for designing media information systems, for example, making typologies of customers.

Not the calculation that we will have learned but the way to understand the problems and solve them. Thus, the field of speech processing, using mathematical methods of signal processing, and appropriate statistical modeling and hidden Markov models compiled into stochastic automata. In addition, we refer to reports on vocational education and the estimation of mathematics programs for the vocational degrees, which may be too ambitious. On the other hand, to provide an example, we may also use computer science, mathematics, and mathematical logic for a concept of language. A cooking recipe, an algorithm to cross the street, can be executed and even transmitted without being written or verbalized.

This would avoid a juxtaposition of unrealistic texts. This is the question: What is the lower bound on the cost of any calculation algorithm that responds to a given problem? Note the problem P =NP, to provide an example.

The field of discrete randomness is approached nowadays mainly from complementary but not contradictory angles. Thus, the first applications are largely the result of digital analysis of differential systems. They also suggest that the best way to train future generations of high-performance computer scientists is to provide them with training adapted to the concepts, objects, and methods of computer science. Another way of approaching this problem is to extend the verification of programs, traditionally carried out by a battery of tests, into a symbolic verification in a more abstract computing space-called model checking.

This notion of non-deterministic computation verifies remarkable properties, notably confluence, which expresses that the calculations converge into a single normal form. This is to ensure confidentiality and authenticity of communications on networks open to all. Finally, with the emergence of Big Data, data is becoming an essential commodity for many organizations and businesses.

To do this, engineering colleges remain the primary road. In the end, all we do in computing is math. It is interesting to note that these fundamental trends still constitute the basis of contemporary computing. It is a challenge to identify a set of general methods and to find what might be the analog of conservation laws in physics or the second principle of thermodynamics for computation.

On the other hand, the reduction of polynomial systems is related to contemporary algebraic geometry, the symbolic manipulation of many special functions of physics becomes possible thanks to the joint use of modern differential algebra and adapted data representations. Polynomial factorization is best done by methods that exploit both the structure of the finite fields and the principles of randomization.

Finally, I'll finish with a nuance: a computer language is a formal way of communicating with a machine, and is therefore essentially a mathematical tool as the machine is based on logical principles.

BIBLIOGRAPHY

1. Angluin, D., & Valiant, L. (1979). Probabilistic Algorithms for Hamiltonian Circuits and Matchings. *J. Comput. Sys. Sci. 18*, 155–190.
2. Applegate, D., Bixby, R., Chvatal, V., & Cook, W. *Solving Traveling Salesman Problems*. http://www.tsp.gatech.edu/.
3. Applegate, D., Bixby, R., Chvatal, V., & Cook, W. (1995). Finding Cuts in the TSP (a Preliminary Report). *Technical Report 95-05, DIMACS*. Piscataway NJ: Rutgers University.
4. Bollobás, B. (1979). *Graph Theory: An Introductory Course*. New York: Springer-Verlag, p. 12.
5. Borwein, J., & Bailey, D. (2003). *Mathematics by Experiment: Plausible Reasoning in the 21st Century.* Wellesley, MA: A K Peters, pp. 4–5.
6. Borwein, J. M., Borwein, P. B. Pi., & the AGM. (1987). *A Study in Analytic Number Theory and Computational Complexity*. New York: Wiley.
7. Bridges, D. S. (1994). *Computability*. New York: Springer-Verlag.
8. Brookshear, J. G. (1989). *Theory of Computation: Formal Languages, Automata, and Complexity*. Redwood City, CA: Benjamin/Cummings.
9. Chalaturnyk, A. (2008). *A Fast Algorithm for Finding Hamilton Cycles*. Master's thesis. Winnipeg, Manitoba, Canada: University of Manitobaftp://www.combinatorialmath.ca/g&g/chalaturnykthesis.pdf.
10. Chartrand, G. (1985). *Introductory Graph Theory*. New York: Dover, p. 68.

11. Clay Mathematics Institute. P vs. NP. http://www.claymath.org/millennium/P_vs_NP/.Cook, S. The P versus NP Problem. http://www.claymath.org/millennium/P_vs_NP/Official_Problem_Description.pdf.
12. Cooper, S. B., Slaman, T. A., & Wainer, S. S. (Eds.). (1996). Computability, enumerability, unsolvability: directions in recursion theory. New York: Cambridge University Press.
13. Crescenzi, P., & Kann, V. (Eds.). *A Compendium of NP Optimization Problems*. http://www.nada.kth.se/~viggo/wwwcompendium/wwwcompendium.html.
14. Csehi, C. Gy., & Tóth, J. (2011). Search for Hamiltonian Cycles. Mathematica J. 13http://www.mathematica-journal.com/2011/05/search-for-hamiltonian-cycles/.
15. Davis, M. (1982). *Computability and Unsolvability*. New York: Dover.
16. Du, D. Z., & Ko, K. I. (2000). *Theory of Computational Complexity*. New York; Wiley.
17. Gardner, M. (1957). Mathematical games: about the remarkable similarity between the Icosian game and the towers of Hanoi. *Sci. Amer. 196*, 150–156.
18. Gardner, M. (1984). *The Sixth Book of Mathematical Games from Scientific American*. Chicago, IL: University of Chicago Press, pp. 96–97.
19. Gardner, M. (1986). The Binary Gray Code. In *Knotted Doughnuts and Other Mathematical Entertainments*. New York: W. H. Freeman, pp. 23–24.
20. Garey, M. R., & Johnson, D. S. (1983). *Computers and Intractability: A Guide to the Theory of NP-Completeness*. New York: W. H. Freeman.
21. Goetz, P. (2018). Phil Goetz's Complexity Dictionary. http://www.cs.buffalo.edu/~goetz/dict.html.
22. Greenlaw, R., Hoover, H. J., & Ruzzo, W. L. (1995). *Limits to Parallel Computation: P-Completeness Theory.* Oxford, England: Oxford University Press.
23. Griffor, E. R. (Ed.). (1999). *Handbook of Computability Theory.* Amsterdam, Netherlands: Elsevier.
24. Hoffman, P. (1998). *The Man Who Loved Only Numbers: The Story of Paul Erdős and the Search for Mathematical Truth*. New York: Hyperion, pp. 168–169.

25. Hopcroft, J. E., & Ullman, J. D. (1979). *Introduction to Automated Theory, Languages, and Computation.* Reading, MA: Addison-Wesley.

26. Karp, R. M. (1972). Reducibility Among Combinatorial Problems. *In Complexity of Computer Computations*, Miller, R. E. & Thatcher, J. W. (Eds.). New York: Plenum Press, pp. 85–103.

27. Khomenko, N. P., & Golovko, L. D. (1972). Identifying Certain Types of Parts of a Graph and Computing Their Number. *Ukr. Math. J. 24*, 313–321.

28. Kocay, W. (1992). An Extension of the Multi-Path Algorithm for Hamilton Cycles. *Disc. Math. 101*, 171–188

29. Kocay, W., & Li, B. (1994). *An Algorithm for Finding a Long Path in a Graph*. Util. Math. *45*, 169–185

30. Kruskal, J. B. (1956). On the Shortest Spanning Subtree of a Graph and the Traveling Salesman Problem. *Proc. Amer. Math. Soc. 7*, 48–50

31. Lawler, E., Lenstra, J., Rinnooy Kan, A., & Shmoys, D. (1985). The Traveling Salesman Problem: a Guided Tour of Combinatorial Optimization. New York: Wiley.

32. Lederberg, J. (1967). Hamilton Circuits of Convex Trivalent Polyhedra (up to 18 Vertices). *Amer. Math. Monthly 74*, 522–527.

33. Lewis, H. R., & Papadimitriou, C. H. (1997). *Elements of the Theory of Computation*, 2nd ed. Englewood Cliffs, NJ: Prentice-Hall.

34. Lin, S. (1965). *Computer Solutions of the Traveling Salesman Problem.* Bell System Tech. J. 44, 2245–2269.

35. Ore, O. (1960). A Note on Hamiltonian Circuits. *Amer. Math. Monthly 67*, 55.

36. Perepechko, S. N., & Voropaev, A. N. The number of fixed length cycles in an undirected graph. Explicit Formulae in Case of Small Lengths.

37. Platzman, L. K., & Bartholdi, J. J. (1989). Space-filling Curves and the Planar Travelling Salesman Problem. *J. Assoc. Comput.* Mach. *46*, 719–737.

38. Reinelt, G. (1991). TSPLIB--A Traveling Salesman Problem Library. *ORSA J. Comput. 3*, 376–384.

39. Rosenkrantz, D. J., Stearns, R. E., & Lewis, P. M. (1977). An Analysis of Several Heuristics for the Traveling Salesman Problem. *SIAM J. Comput. 6*, 563–581.

40. Rubin, F. (1974). *A Search Procedure for Hamilton Paths and Circuits. J. ACM 21*, 576–580.
41. Skiena, S. (1990). Traveling Salesman Tours. 5.3.5 in *Implementing Discrete Mathematics: Combinatorics and Graph Theory with Mathematica*. Reading, MA: Addison-Wesley, pp. 199–202.
42. Skiena, S. (1990). *Implementing Discrete Mathematics: Combinatorics and Graph Theory with Mathematica*. Reading, MA: Addison-Wesley.
43. Skiena, S. (1990). Hamiltonian Cycles. 5.3.4 in Implementing Discrete Mathematics: Combinatorics and Graph Theory with Mathematica. Reading, MA: Addison-Wesley, pp. 196–198.
44. Skiena, S. S. (1997). Traveling Salesman Problem. 8.5.4 in *The Algorithm Design Manual*. New York: Springer-Verlag, pp. 319–322.
45. Sloane, N. J. A. (2018). Sequences A003042/M2053, A005843/M0985, A006069/M1903, A007395/M0208, A094047, A124349, A124353, A124355, A124356, A129348, A129349, A143246, A143247, A143248, A143249, A174589, A280847, and A281255 in The Online Encyclopedia of Integer Sequences.
46. Smale, S. (1998). Mathematical Problems for the Next Century. *Math. Intelligencer 20*(2), 7–15.
47. Smale, S. (2000). Mathematical Problems for the Next Century. *In Mathematics: Frontiers and Perspectives 2000,* Arnold, V., Atiyah, M., Lax, P., & Mazur, B. (Eds.). Providence, RI: Amer. Math. Soc.
48. Steinhaus, H. (1999). *Mathematical Snapshots*, 3rd edn. New York: Dover, pp. 120–121.
49. Sudkamp, T. A. (1996). *Language and Machines: An Introduction to the Theory of Computer Science*, 2nd edn. Reading, MA: Addison-Wesley.
50. Tutte, W. T. (1946). On Hamiltonian Circuits. *J. London Math. Soc. 21*, 98–101.
51. Vandegriend, B. (1998). *Finding Hamiltonian Cycles: Algorithms, Graphs and Performance. Master's thesis, Winnipeg, Manitoba,* Canada: University of Manitoba.
52. Weisstein, E. W. Books about Computational Complexity. http://www.ericweisstein.com/encyclopedias/books/ComputationalComplexity.html.
53. Welsh, D. J. A. (1993). *Complexity: Knots, Colorings and Counting*. New York: Cambridge University Press.

54. Wilf, H., & Algorithms, S., (1994). *Complexity*. pp. 120–122. Summer. http://www.math.upenn.edu/~wilf/AlgoComp.pdf.

INDEX

D

E

F

G

H

I

L

M

N

O

P

R

S